Getting into University
Engineering Courses

James Bourne

8th edition

trotman | t

Getting into University: Engineering Courses

This 8th edition published in 2025 by Trotman, an imprint of Trotman Indigo Publishing Ltd, 18e Charles Street, Bath BA1 1HX

© Trotman Indigo Publishing Ltd 2025

Author: James Bourne
7th edn: James Barton
6th edn: James Barton and John Southworth
1st–5th edns: James Burnett

British Library Cataloguing in Publication Data
A catalogue record for this book is available from the British Library.

Paperback ISBN 978 1 911724 42 1
eISBN 978 1 911724 43 8

All rights reserved. This book is sold subject to the condition that it shall not, by way of trade or otherwise, be lent, resold, hired out or otherwise circulated without the publisher's prior written consent in any form of binding or cover other than that in which it is published and without a similar condition including this condition being imposed on the subsequent purchaser. No part of this publication may be reproduced, stored in a retrieval system or transmitted in any form or by any means, electronic and mechanical, photocopying, recording or otherwise without prior permission of Trotman Indigo Publishing.

Every effort has been made to trace copyright holders and to obtain their permission for the use of copyright material. The publisher apologises for any errors or omissions, and would be grateful to be notified of any corrections that should be incorporated in future editions of this book.

The authorised representative in the EEA is Easy Access System Europe Oü (EAS), Mustamäe tee 50, 10621 Tallinn, Estonia.

Printed and bound in the UK by 4Edge Ltd, Hockley, Essex

 All details in this book were correct at the time of going to press. To keep up to date with all the latest news and updates and to access the online resources that accompany this book, visit www.trotman.co.uk/pages/getting-into-online-resources

Contents

About the author	vi
Acknowledgements	vii
Re-inventing the wheel: Introduction	**1**
About this book	4
Engineering defined	4
Opportunities for engineers	10
Sixth form considerations	11
Competition for places	13
1 The nuts and bolts: Studying engineering	**15**
Engineering courses	15
Project work	20
The importance of mathematics	20
Methods of assessment and study	21
BEng v. MEng courses	22
Combined honours courses	22
Other courses	23
2 Building blocks: Getting work experience	**26**
Looking for work experience	27
A sample CV	29
The covering letter	31
Work experience interviews	32
How to use your work experience effectively	32
3 Measure twice, cut once: Choosing your course	**34**
What to consider	34
League tables	37
Choosing the right course	38
Overseas study	40
Academic and career-related factors	41
Non-academic considerations	43
4 Righty tighty, lefty loosey: Completing your UCAS application	**46**
The UCAS application	46
Suggested timescale	47
Entrance examinations	50
Taking a gap year: Deferred entry	51

5	**Fire on all cylinders: The personal statement**	**56**
	Who will read your personal statement?	56
	Joint honours courses	57
	Applying for different courses at the same university	57
	The structure of the personal statement	58
	Personal statement writing with Artificial Intelligence (AI)	60
	Analysing a personal statement	63
	Linking your interests and experiences	67
	Work experience	67
	How to get started on the personal statement	68
	Language	70
	Sample personal statements	72
6	**A well-oiled machine: Succeeding at interview**	**78**
	Preparing for an interview	80
	The interview	82
	How to answer interview questions	87
	Steering the interview	90
	Current issues	91
7	**Don't get your wires crossed: Non-standard applications**	**105**
	Changing direction	105
	Mature students	106
	International students	108
	Students with disabilities and special educational needs	111
8	**Elbow grease: Results day**	**113**
	When the results are available	113
	What to do if you have no offers: UCAS Extra	113
	What to do if things go wrong during the exams	114
	What to do if you exceeded the grades that you expected	115
	What to do if you have no confirmed offers	116
	UCAS Clearing – Tips	117
	Advice for Clearing	117
	If you decide to retake your A levels	118
	If you decide to reapply	118
9	**Building bridges: Fees and funding**	**120**
	Fees	120
	Living expenses	121
	Funding your studies	122
	Tuition fee loans	122
	Maintenance loans	123
	Sponsorship	125
	Scholarships	125
	Special considerations	126

10	**A cog in the machine: Further training, qualifications and careers**	**128**
	Chartered Engineer status	130
	Incorporated Engineer status	131
	Master's courses	132
	Career opportunities and employment prospects	133
11	**Light years ahead: Further information**	**135**
	Useful contacts	135
	Specialist publications	137
	Books	138
	Glossary	**140**

About the author

James Bourne is the Vice Principal of MPW Birmingham, having taught at the college since 2006. Holding a Master of Arts (MA) in Illustration, James has a wealth of experience in supporting students through the UCAS application process. His areas of expertise include both the arts and engineering, and he has guided numerous students, helping them navigate the complexities of university admissions and make informed decisions about their academic futures.

Acknowledgements

I would like to express my gratitude to the authors of previous editions of *Getting into University: Engineering Courses*, including James Burnett, James Barton and John Southworth, whose contributions have laid a solid foundation for this guide.

I am appreciative of the former MPW students whose insights and experiences have been invaluable in summarising the perspectives of both undergraduate and postgraduate engineering students.

A special thanks goes to the staff at MPW Birmingham, whose extensive knowledge and expertise in guiding students through the undergraduate application process have been instrumental in shaping the core content of this updated edition. Their unwavering commitment to student success continues to provide a vital resource for prospective applicants.

Re-inventing the wheel: Introduction

> *Scientists study the world as it is; engineers create the world that has never been.*
>
> *– Theodore von Kármán*

To an extent, Theodore von Kármán's observation is accurate; engineering is not just a tool but a force for creating possibilities that never existed before. We owe much to engineering for shaping and transforming our lives since the dawn of humanity. As von Kármán noted, while science helps us understand the world as it is, engineering empowers us to innovate and build a future that meets our evolving needs, ensuring we can adapt and thrive in the face of challenges ahead.

Engineering is the past, the present and the future. Ask anyone what the greatest engineering ideas in history are, and you will get different answers. Stonehenge is arguably a good example to begin with, although modern engineering has not yet allowed a bypass to be built there yet. The Egyptians, the Greeks, the Incas and the Romans were all engineering innovators. They continue to set examples; why else would the Crown Prince of Saudi Arabia announce his wish to create 'the modern Pyramids' in Neom, his 2030 vision for the Kingdom of Saudi Arabia: a 100-mile city, 200 metres wide, built to be 100% sustainable and with 95% of land given to nature; it would be an unbelievable feat of engineering.

Closer to home, Birmingham stands as a testament to the transformative power of engineering and design, with projects such as the HS2 railway and the regeneration of areas like Digbeth shaping the city's future. Further afield, the railway line from Brighton into London passes through East Croydon, an area currently undergoing significant redevelopment. As you arrive in London, you pass Battersea Power Station, still in the midst of its transformation into high-end flats and a now-open shopping centre. When crossing the River Thames, the change London has experienced throughout history is most strikingly visible in the engineering projects along the riverbanks. From buildings dating back to Queen Victoria's Great Exhibition to the 21st-century Shard, the feats of structural and civil engineering are impressive, inspiring, diverse and challenging. In so many ways, design and engineering are intertwined;

the responsibility of any architect and engineer is to ensure that these structures fit within the legacy of the city.

The bicycle, computers, 3D printers – the characteristic feature of our world is that it will continue to adapt, and engineers are at the forefront of that. Whether we are talking about new building projects, drones, 5G infrastructure, Virtual Reality (VR), roads, railways, Formula One on the cusp of changing to Formula E in the next 10 years, commercial flights to space, new super liners, the Hadron Collider or just basic amenities, there is always something to be said about engineering. With the growing influence of Artificial Intelligence (AI), engineers now have unprecedented tools to design, optimise and innovate from AI-driven construction techniques to machine learning models revolutionising aerospace and automotive design. The profession is not often seen as having any glamour, and yet the exciting thing about a career in this sector is that it can be whatever you want it to be.

Few things will happen in our lives without the influence of engineering in some respect or other, yet the sector has never had as high a profile as, for instance, architecture, medicine or economics among the public. However, in recent years that has begun to change, as engineering projects ranging in scale from nanotechnology and the electronics that create your smartphone to amazing feats such as sending the Rosetta spacecraft to Comet 67P have captured the public's imagination. Engineers are also involved in environmental and safety issues, as well as helping to shape the world's future energy and raw materials needs.

In the modern age, sustainability is the word on everyone's lips, and carbon neutral is a mantra for businesses to follow. For example, Poseidon is leading the way on recyclable plastics, turning waste into 100% recyclable plastic. This is a fascinating project for upcoming engineers to follow closely, as it is a prime example of how engineering crosses codes – chemical, mechanical and systems-based engineering.

Engineering projects are also being introduced to young children in schools: the Barrow Engineering Project at the Royal Academy of Engineering, for example, gives students in Barrow-in-Furness the chance to learn about engineering from engineers and technicians. Projects such as these are essential to help the profession attract new audiences year on year.

If you are interested in engineering, you might be able to name some famous engineers – Thomas Edison, Isambard Kingdom Brunel, James Dyson, George Stephenson and Tim Berners-Lee, for example – but most engineers work as part of teams; teams that work on making our lives as fulfilling, safe and manageable as possible without people being aware of what they do. There would be no exciting new buildings or bridges without the structural engineers to work alongside the architects; no iPhones or iPads without the electronic, telecommunication and materials engineers; no websites without software engineers; and Rosetta would not have started its mission without the mechanical, chemical and aerospace engineers who created it and successfully launched it from Earth.

As you are reading this introduction, you are probably thinking about what opportunities exist for engineers. I hope that reading this book, coupled with your research into universities and careers within engineering, will show you that there is a myriad of exciting possibilities waiting for you within this field. The most thrilling part of becoming an engineer is stepping into the unknown and being part of a future where innovations yet to be imagined will shape the world. Whether it's breakthroughs in sustainable energy, space exploration, AI or technologies we can't yet conceive, you have the chance to contribute to groundbreaking developments and leave your mark on history.

Chapter 1 focuses on what studying engineering at university entails, including the structure of degree courses, the method of assessment and the different options for course length and type.

Chapter 2 looks at the desirability of gaining work experience or participating in taster courses and how to go about finding suitable placements.

Chapter 3 covers the process of how to choose the right engineering course for you. There is information on how to start your search and what factors (academic and non-academic) you should use to filter your choices in order to end up with the five courses for the UCAS application.

Chapters 4 and **5** give you the essential information that you need when using the UCAS system to apply and writing the all-important UCAS personal statement.

Some universities require students to attend an interview so **Chapter 6** discusses how to prepare for interviews and what to expect.

Chapter 7 contains advice for students who decide to study engineering as a change of direction, mature applicants, students with disabilities and those who are applying from outside the UK.

In **Chapter 8**, there is a breakdown of the options available when you get your examination results and what to do if you do not achieve the grades you require.

Financial information, both the cost of studying and what financial support may be available, is covered in **Chapter 9**.

Chapter 10 looks at options for further study and training for graduates and includes information on career opportunities and Chartered Engineer status.

Chapter 11 lists sources of further information for potential engineers, and at the end, there is a **Glossary** of common terms used in this book.

About this book

The aim of this book is to take you through the process of applying to study engineering, from choosing courses and universities through to postgraduate courses and career opportunities.

You can use this book as a source of advice and information by reading the chapters or sections that are of relevance to you. However, I recommend that you read the book from the beginning rather than dipping in and out of it, because you will then get a more complete picture.

References to university entrance requirements throughout the book are usually given in terms of A level grades, and the equivalent entry requirements for students studying the International Baccalaureate (IB), Scottish Highers and other qualifications can be found by using the UCAS Tariff (see Chapter 3). Universities, on their websites, provide further details of entrance requirements for all of the commonly accepted examinations, as does the 'Find a course' facility on the UCAS website (www.ucas.com). Regardless of the examination system you are using, the advice on applications given in the book is applicable to all candidates. If you have any questions about your own particular situation, universities are happy to deal with individual queries and can be contacted via their 'contact us' sections on their websites.

Engineering defined

What is engineering?

What is the difference between science and engineering? There are many definitions, but, essentially, engineering is the practical application of mathematics and science to create machines, processes or structures. Whereas the starting point in science generally involves trying to explain or predict phenomena through the development and verification of theories and models, engineering is the process of physically achieving a goal by applying scientific ideas and theories in a practical way.

There are many different types of engineering, all of which have a large impact on our lives. Good examples of this include bridge design to the redevelopment of cities, from Formula One to prosthetics – though a specific one for today would be the ongoing HS2 line, set to connect London to Birmingham in its first phase. Originally intended to extend to the north of England, the northern phases were cancelled in 2023.

The first stage is expected to become operational between 2029 and 2033. Like many major engineering projects, HS2 has faced challenges, including delays and budget overruns, now estimated at £49–£56.6 billion. Another example of extraordinary engineering is the Elizabeth line in London. Adding a new tube line through the centre of London was an enormous undertaking and a feat of structural and civil engineering. From constructing the 21-kilometre twin-bored tunnels to performing calculations to ensure no ground movement affects structures above, the project showcases the interdisciplinary skills of the profession, combining physics, mathematics, design and construction.

What is an engineer?

Dreamer. Innovator. Researcher. Problem Solver. Inventor. Creator.
www.whatisengineering.com

What do engineers do?

Although we tend to classify engineers and engineering courses under different headings, there is a good deal of overlap, and most engineering projects or processes require the input of many different types of engineers. When you look in more detail at the course content of different engineering programmes, you will see that there are many common elements to these. For instance, mechanical engineers will spend time studying electronic and electrical engineering, as machines often use electricity as a power source; and civil engineering requires an understanding of the properties of materials, as does structural engineering.

Most engineering projects involve the input of a range of specialisms (see Figures 1 and 2 overleaf).

Aerospace engineers
Aerospace engineering is concerned with the design of aircraft and spacecraft. It is made up of aeronautical engineering and astronautical engineering. The job of the engineer is concerned with the design and propulsion of the aircraft, as well as reviewing the effectiveness of the craft and the materials from which it has been constructed.

Agricultural engineers
Agricultural engineering focuses on everything from agricultural equipment and machinery through to bioenergy and development of resources on farms.

AI and machine learning engineers
AI and machine learning engineering involves creating systems capable of intelligent behaviour by developing algorithms that enable machines to learn from data. These engineers design applications such as autonomous vehicles, recommendation systems, speech recognition tools and predictive analytics. AI engineers often work at

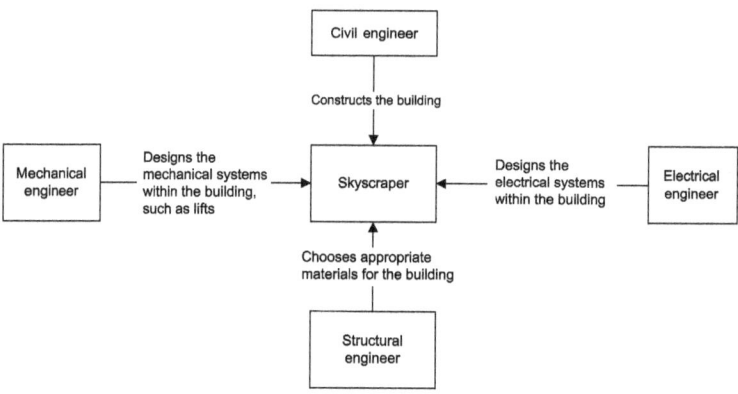

Figure 1 Input into building a skyscraper

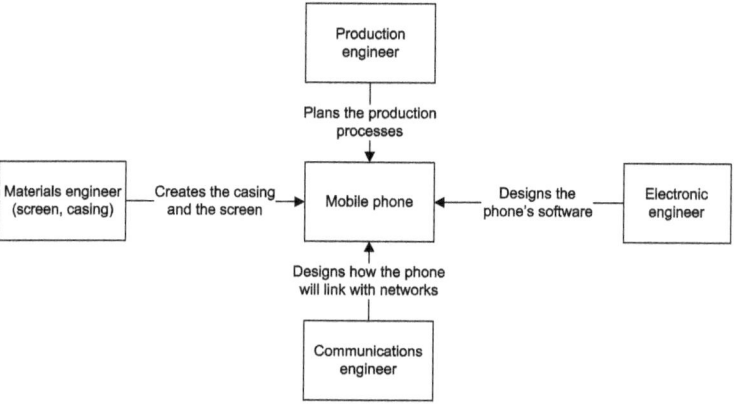

Figure 2 Input into building a mobile phone

the intersection of software engineering and data science, applying advanced mathematics and computational techniques to solve problems. As AI increasingly shapes industries such as healthcare, finance and entertainment, the role of these engineers continues to expand in importance.

Biomedical engineers

Biomedical engineering (along with biotechnology) looks at the engineering aspects of living things, often for medical purposes. This can range from working with living materials, such as animal tissue, to the development of medical instrumentation (medical scanners and equipment used in surgery) and the design of machines and devices such as heart pacemakers and artificial limbs.

Chemical engineers

Chemical engineering deals with the industrial processes that produce, for example, drugs, food and fuels. Chemical engineers are concerned with not only the chemical properties of the materials they are producing or developing but also the economic and safety aspects of the projects. Chemical engineering links closely with bioengineering, biomedical engineering and biotechnology.

Civil engineers

Civil engineering deals with the large-scale infrastructure that is an essential part of daily life, such as roads, bridges, dams, water supplies, and office and apartment blocks. (The name civil engineering came about in order to differentiate projects that were there to benefit society in general from military engineering projects.)

Computer engineers

Computer engineers design all sorts of technology, from computers to mobile devices, embedded computer systems to machine learning and AI. It was computer engineers who brought about the digital age, and they are the reason we have infrastructure capable of handling the security risks these advances pose.

Design and product engineers

Design and product engineering deals with the process of creating and developing systems and devices. As well as looking at production processes, such as the design of production lines and factories, design and product engineers have to be aware of safety and cost issues. As an example, the production of a new car involves mechanical and electrical engineering input, computer hardware and software development, the choice of the right materials to ensure that the car will be safe, functional and attractive, and the creation of a production line that will be both efficient and cost-effective.

Electrical and electronic engineers

Electrical and electronic engineers work with electrical and electronic devices. As with mechanical engineering, these range from a microscopic scale (e.g. integrated circuits or solid-state devices) through to national electricity networks. Many universities offer a joint electrical/electronic engineering course, but it is also possible to specialise in just one of these subjects. There are close links between electronic engineering and computer or information technology (IT) engineering.

Energy engineers

Focusing on the energy industry, energy engineers work on electricity generation and transmission with an emphasis on alternative sources of energy, energy efficiency and environmental issues. Energy engineering is a rapidly growing field, encompassing traditional (fossil fuels, nuclear fission) and alternative (wind turbines, wave and tidal schemes, solar power, biofuels) sources of energy.

Environmental engineers

This field is solely focused on the protection and improvement of the environment and the world around us, including the removal of plastic waste, for instance.

Information systems engineers

Information systems engineering is closely linked to electronic engineering. It focuses on computer systems and the transfer of electronic and digital information – mobile phones, the internet and computer operating systems.

Marine engineers

Marine engineers look after anything at sea, including boats, oil rigs, ships, submarines and so on. There is a need for engineering sciences and design principles in all aspects of shipping, drilling and transportation above and below the water, including vessels, pipes, rigging, cables and so on. With bigger and bigger cruise ships being built, more and more naval vessels being commissioned, and more work on wind farms at sea, this form of engineering will endure into the future.

Materials engineers

Materials engineering is closely related to structural engineering in that it looks at the properties of materials that are necessary to create structures. However, it also covers the use of materials for other requirements, such as plastics, ceramics, glass and polymers. A mobile phone manufacturer, for example, may work with materials engineers to ensure that the phone is strong enough to withstand daily knocks while at the same time being light and attractive.

Mechanical engineers

Mechanical engineers work in the development and manufacture of machines. This is obviously a very broad description: the word 'machine' covers an enormous range of devices, from medical equipment that is used to perform microsurgery to aircraft carriers. Mechanical engineering courses include a number of specialist areas, such as aeronautical engineering and automotive engineering. Aeronautical engineers work on all aspects of aircraft and spacecraft, from aerodynamics to the design of engines. Automotive engineers can also work on things that fly, as well as cars and other forms of transport.

Petroleum engineers
Petroleum engineering covers all aspects of the oil, gas and petroleum industries – from exploration and excavation through refining and purification to distribution. It covers geological studies, the chemical properties of hydrocarbons and industrial processes.

Process (systems) engineers
Chemical, mining, petroleum, biotechnology and pharmaceutical engineering can all be described as being branches of process engineering. Process engineers use chemical and biological processes to make raw materials into useful commodities. Process systems engineers tend to work on IT and computer programmes to facilitate these processes.

Robotic engineers
Robotic engineering focuses on the design, development and programming of robots and automated systems that perform tasks in varied settings. These engineers combine expertise in mechanical engineering, electronics and computer science to create robots capable of activities ranging from assembling products in factories to performing surgeries with ultraprecision. Modern advancements have expanded the field to include autonomous drones, humanoid robots and machines used in hazardous environments such as deep-sea exploration or disaster recovery. The development of soft robotics and AI integration has further propelled robotics into areas like healthcare, agriculture and space exploration, making this an ever-evolving field.

Sports engineers
This branch of engineering is concerned with the development, design and testing of sports equipment to improve the game and the safety of the players; this is partly corporately driven.

Structural engineers
Structural engineering deals with the use and suitability of materials intended for creating structures such as buildings, bridges, stadiums and electricity pylons. The discipline covers the structure and properties of materials on a microscopic level and the behaviour of structures on a macroscopic scale. Structural engineers work closely with architects and civil engineers. Designing a breathtaking structure based on an architect's sketch can be a creative and rewarding process, as can handing it over to the civil engineers to build. The career can also be attractive to those who like using modern computer-aided design tools.

Textile engineers
Textile engineers are responsible for creating new fabrics and fibres concerning anything from clothes to papers and cardboard products.

There are many other classifications of engineering disciplines and courses, and there are subsets of some of the areas mentioned above and courses that combine two or more of these disciplines. So, you will need to spend some time researching possible courses before deciding what you want to do. Advice on this can be found in Chapter 3.

Opportunities for engineers

Very few careers provide as many opportunities as engineering while at the same time offering secure employment prospects.

- Engineers can work anywhere in the world.
- The work can be theoretical or practical and can be carried out within an office or on-site.
- Engineers can work for multinational companies or set up their own business.
- Engineers can work on any scale, from nanotechnology and microelectronics through to building the world's biggest structures.
- The world will always need engineers.

Case studies

Case studies are an interesting and informative way to find out more about engineering. Most university websites will have case studies of graduates, and the professional engineering bodies (see Chapter 11) illustrate their particular disciplines using case studies. The Royal Academy of Engineering's website has a good selection and would make an ideal starting point for those researching a career in engineering: www.raeng.org.uk/education/this-is-engineering.

> **Case study**
>
> 'I grew up watching the great, late Stirling Moss achieve what he did in the golden age of motor cars. It really inspired me and drove (forgive the pun) me to want to be part of that world. However, I was never really that interested in becoming a driver. I wanted to be responsible for what made these magnificent machines work.
>
> 'So I went and studied Mechanical Engineering at the University of Bath. It was an excellent course, but the path I wanted to take was not apparent to me at the time. Although a boyhood goal, motorsport at that point did not want me, or at the very least did not want to show me how to access it.

Introduction

> 'Instead, I went to work as a mechanical engineer in the construction industry, which, though not what I had originally envisaged, was where I was really moulded in my early career. It gave me the opportunity to travel and to work on some exceptional projects. After doing this for about five years, I heard about an apprenticeship with Aston Martin. I applied and got offered a place, learning the job alongside some of the best car mechanics of their generation.
>
> 'I learnt a lot at Aston Martin, from the design process to the complexities of the engines. I was lucky enough to work on the development of the DB model and the V6 to V8 engines. I spent a very happy 12 years with the company before finally moving into my dream role. I remember meeting the Williams team boss at an event one day, and we got to discussing cars. From there, I received a call to interview for a role in the Formula 2 team, which I got and stayed with the team for eight years before moving into the Formula 1 paddock in a technical role.
>
> 'My story is my own, but if it comes to offering advice to any aspiring engineers out there I would say this: it is great to have dreams; however, never lose sight of the fact you might have to take a roundabout route to get there. This is natural and means we are shaped by our experiences. I certainly became a better motorsport engineer from taking a more general mechanical engineering degree because my knowledge was broader.'
>
> Maurice Byfield MEng

Working internationally

Few careers provide greater opportunities to work or study overseas, and this is one of the attractions of engineering. Many of the world's major engineering projects are undertaken by multinational companies, and many engineers work freelance, choosing their projects and locations to suit their skills and circumstances. The rapid technological advances of the BRICS nations (Brazil, Russia, India, China and South Africa) have created enormous demands for qualified engineers.

Sixth form considerations

Study combination

Generally speaking, it is considered that the majority of engineering courses will require Mathematics and Physics A levels. There are additional variables to this in the cases of Biomedical Engineering and Chemical Engineering, for instance. The bigger question is what should the third subject be, and what is the need for further mathematics? The best advice for the third subject is to take something that you are going to enjoy, as the grade will talk at the end of the day. That is not to say go off-piste in your choices; it is important to remember what it is that you are gaining from that qualification. A lot of people question the requirement for further mathematics at sixth form. The answer to that is ambiguous. It is not usually required by a university, unless overtly stated, because there are schools that don't offer it as a subject. That said, the wording can often imply that it would be preferred if you had it. The simple reading is that, if you can, you might wish to consider it because you will be studying further mathematics as part of an engineering course anyway, but it will not break your application if you do not have it.

T levels

Launched in September 2020, a T level is a two-year qualification equivalent to three A levels. They have been developed alongside guidance from employers and businesses and aim to provide students with a platform from which they can develop skill sets for specific industries. The learning is balanced between 80% classroom teaching and 20% on-site experience, for which students need to complete at least 315 hours of work (45 days). This route is increasingly recognised by universities and employers, so students pursuing it can confidently progress to higher education or a career in engineering.

There are three distinct engineering pathways:

- Engineering, processing, manufacturing and control;
- Designing, surveying and planning for construction (civil engineering);
- Building services engineering for construction.

For the grading and UCAS equivalent points, see Chapter 3.

EPQ

The EPQ enables students to develop and demonstrate particular interests outside of their core A level subjects. The EPQ is worth 50% of an A level (in terms of UCAS points) and can add value to Russell Group and/or Oxbridge applications.

> We welcome the Extended Project and would encourage applicants to undertake one as it will help to develop independent study and research skills valuable for higher education.
> University of Cambridge

Typical university offers to read BEng Engineering are AAA, AAB with an A at EPQ or ABB with an A* at EPQ. The EPQ is not taught; rather, students work independently to research and develop their project. In conjunction with their supervisor, the student chooses a topic and a format for the project. This can be a research-based essay (of between

4,000 and 5,000 words), a creative piece (such as a short play or short story), an artefact (such as a model) or a presentation.

Competition for places

While the ratio of applicants to places for engineering courses is lower than for many other subjects, the competition for places at the higher-ranked universities is intense. Thus, many candidates, while being successful at gaining a place on an engineering degree course, do so either through Clearing or at one of their lower-preference universities. You should aim as high as you can (within the boundaries set by your examination results and predictions) in terms of your choice of university, as employers will look not only at what you studied but also at where you studied. Therefore, do all that you can to ensure that your choice of university will stand you in good stead in the future.

At the time of writing, applications to universities continue to rise. This does not necessarily mean there are more places, as the top universities are still constrained by space and laboratory/workshop facilities; there are still far more applicants than places at these institutions. In 2023, for engineering courses overall, there were 189,495 applicants, for 32,275 places, with the most popular areas being mechanical engineering (7,675), electrical and electronic engineering (4,955) and general engineering (3,960). Of the 32,275 successful applicants 25,620 were men and 6,655 were women. As Table 1 shows you, the competition for places at universities in the UK across all aspects of engineering is intense, and the picture is fairly constant year on year.

Table 1 Types of engineering courses and application numbers 2022 and 2023

Type of engineering	2022 applications (number of places)	2023 applications (number of places)
General engineering	21,490 (4,115)	21,255 (3,960)
Civil engineering	23,625 (4,040)	22,695 (3,830)
Mechanical engineering	47,720 (7,750)	48,720 (7,675)
Aerospace engineering	23,000 (3,750)	23,870 (3,830)
Naval architecture	195 (35)	240 (45)
Electrical and electronic engineering	30,800 (5,065)	30,550 (4,955)
Production and manufacturing engineering	9,250 (2,250)	8,930 (2,060)
Chemical, process and energy engineering	12,880 (2,280)	14,080 (2,425)
Other engineering	60 (20)	60 (15)

As with most things in life, the more planning and preparation you do, the better your chances of success. This is, of course, true for your studies and examinations, but it is also the case that research and a well-planned UCAS application will give you a much better chance of being made offers by your chosen universities. This applies to all aspects of the application: choosing your courses and universities, looking at their entrance requirements, writing a personal statement that will demonstrate your seriousness about, and suitability for, the course, and ensuring that you are prepared for any interviews or entrance tests.

Ignore those who tell you that 'if you are lucky, you will get an offer' – if you pitch your applications at the level that is appropriate for your qualifications (past and predicted) and prepare properly, you can remove most of the uncertainty from your application. The following chapters will tell you how to do this.

> *I was originally supposed to become an engineer but the thought of having to expend my creative energy on things that make practical everyday life even more refined, with a loathsome capital gain as the goal, was unbearable to me.*
> Albert Einstein, theoretical physicist

1 | The nuts and bolts: Studying engineering

What does studying engineering actually entail? Is it all practical work? How much mathematics is involved? Should I take a joint honours degree? This chapter covers these and many other questions. Once you have read the chapter, my advice is to go to the university websites, as they will provide you with more detail as well as case histories and comments from current and past students.

Engineering courses

The structure of an engineering course will vary not only from discipline to discipline but also from university to university. The common elements to all engineering courses are:

- mathematics (see page 20);
- physical laws (e.g. Newton's laws or the laws of energy conservation or thermodynamics);
- the physics of materials (physical, electrical and thermal properties);
- environmental, safety and health issues;
- cost and other economic issues.

> **The difference between school and degree courses**
>
> 'The students who succeed at university are usually distinguished more by their attitude to independent study, their passion for the subject and their willingness to get involved in other aspects of university life than they are by their entry grades. Engineering students work hard and play hard, and they frequently hold senior positions on student society committees or in volunteering groups.
>
> 'Staff often have to help students to see that bite-sized answers to questions that might have been appropriate at A level need to become well-argued and researched responses. Just because the question is only one sentence doesn't mean that the answer should be.

'A new development across the sector over the last 10 or 20 years has been the recognition that it is not enough to teach engineering students a multitude of technical skills. There is an increasing amount of discussion around professional responsibility, dealing with ethical dilemmas, learning from past disasters, etc. (see Chapter 6), which, if taught well, can be just as stimulating as the technical material. The Royal Academy of Engineering has been helping to pioneer these aspects and its website links to many useful publications in this area. So, a student who has studied mathematics, further mathematics and physics may have to adjust slightly as they realise that the broader skills they learned earlier in their school life will need to be dusted off and brought to bear.'

Lawrence Coates, Professor of Engineering, University of East Anglia

Student experience

'When I was studying for my A levels, I knew what I wanted to do in the future, therefore my combination was easy. For all those considering whether to take further mathematics or not, I can tell you I wish that I had as the crash course at the start of the degree was tough. That said, the subject material was not and I gained a lot from working in small group sizes. University is the biggest eye opener you could ask for. Gone are the support structures from school – if you want help, you have to ask for it; however, the expectation is that you resolve things yourself. And that, at the end of the day, is one of the fundamentals of being an engineer. Having attended a sixth form college, I think I was perfectly ready to be a student and I had those learning skills ingrained into me. The advice I would give anyone is to make the most out of every opportunity and do not fall behind early, this is a degree unlike others in terms of workload.'

Mustafa El-Maghraby, Brunel University London, BEng Automotive Engineering

Student experience

'I am just completing the fourth year of my BEng, during which I spent a year in industry. The course itself has been fascinating though exceptionally tough. There is a good variety in learning conditions and that really does inspire me. I am not someone to enjoy just sitting in lecture theatres, and that is not what the profession is about either. Lectures balance nicely with laboratory work and, of course, my industry placement, which has stood me in good stead as I have been offered a job at the end of this year.

'The grading of the course is a balance between written exams and lab reports and you are very much expected to work as part of a team, something I found very rewarding. That has been the huge appeal of the profession to me.

'Don't be surprised at how quickly the course accelerates beyond your A level knowledge; this can be daunting, but remember that everything worth having should never really be too easy to achieve as, this way, you gain a sense of accomplishment from what you solve.'

*Laura Dehan, University of Sussex,
BEng Mechanical Engineering*

Student experience

'I found the difference between A levels and university to be quite stark. At school, most of the learning takes place under supervised conditions and you are made to do lots of repetition and practice to help you learn. This is not the case in university because the teaching is done in lectures, so the reading, the practice and the learning generally take place without any supervision or direct guidance. The process of learning is much more independent at university because you are expected to know how to learn from your A level years and the staff have less time to answer questions and explain things outside lectures. At university there is often no immediate consequence for not completing work, but you must avoid letting yourself fall behind because you will have to do twice as much work to catch-up!'

*Jordan Massiah, Magdalene College,
University of Cambridge, MEng Engineering*

Look at the course content in detail on the university websites to ensure that your interests are covered. Two examples of the typical range of topics covered as part of a four-year engineering degree (MEng) and a three-year engineering degree (BEng) are shown on the following pages.

Aeronautical and Aerospace Engineering (University of Leeds) MEng

Year 1

- Computers in Engineering Analysis
- Design and Manufacture 1
- Thermofluids 1
- Solid Mechanics
- Engineering Materials
- Engineering Mathematics 1

Year 2

- Engineering Mechanics
- Vibration and Control
- Design and Manufacture 2
- Economics and Management
- Mechatronics and Measurement Systems
- Thermofluids 2

Year 3

- Aerospace Vehicle Design
- Aerodynamics and Aerospace Propulsion
- Aerospace Flight Mechanics
- Individual Engineering Project
- Finite Element Methods of Analysis

Year 4

- Team Project
- Aerospace Structures
- Aerospace Systems Engineering
- Computational Fluid Dynamics Analysis

Source: https://courses.leeds.ac.uk/f414/aeronautical-and-aerospace-engineering-meng-beng#content. Modules are reviewed and are subject to change. Reprinted with kind permission of the University of Leeds.

Aerospace Engineering (University of Sheffield) BEng

Year 1

- Engineering Statics and Dynamics
- Aerospace Aerodynamics and Thermodynamics
- Aerospace Engineering Design, Build and Test

- Introduction to Aerospace Materials
- Analysis and Modelling of Aerospace Systems
- Global Engineering Challenge Week
- Electrical Fundamentals
- Mathematics (Electrical and Aerospace)

Year 2

- Control Engineering
- Embedded Programming
- Aerospace Design II
- Aerodynamics and Heat Transfer
- Aerospace Structures and Dynamics
- Aerospace Materials
- Electrical Energy
- Engineering – You're Hired
- Mathematics II

Year 3

- Aircraft Dynamics and Control
- Aerodynamic Design
- Aero Propulsion
- Aircraft Design
- Managing Engineering Projects and Teams
- Accounting and Law for Engineers
- Aerospace Individual Investigative Project

Choose from:
Aeromechanics optional modules:

- Aerospace Metals
- Advanced Engineering Thermodynamic Cycles
- Manufacturing Systems
- Finite Element Techniques
- Computational Fluid Dynamics

Avionic Systems Stream optional modules:

- State-Space Control Design
- Space Systems Engineering
- Machine Learning
- Aerospace Electrical Power Systems
- Antennas, Radar and Navigation

Source: www.sheffield.ac.uk/undergraduate/courses/2025/aerospace-engineering-beng#modules.
Modules are reviewed and subject to change.
Reprinted with kind permission of the University of Sheffield.

Project work

Project work forms an important part of the later years of a degree course and is arguably the most exciting part of the course. The differences between coursework and projects at school and at university are illustrated below.

A level or equivalent	Degree level
Your teacher helps you because he or she is judged on your results.	Your lecturer provides you with a framework within which the coursework is completed.
Study guides are consulted and marking guidance is utilised.	Some constraints are explained in the handouts.
Marks are maximised by ticking all the right boxes.	You interpret them and add value or originality.
Basically, your teacher tells you what to do.	Basically, your lecturer never tells you what to do.
Generally, there is one right way to do it.	Generally, there are multiple right ways to do it.

The importance of mathematics

All engineers use mathematical methods as an integral part of their work. While much of engineering relies on physical processes to develop and produce devices, machines, structures, fuels or chemicals, these are all underpinned by mathematical calculations and models. If you look at the course outlines for engineering degrees, you will notice that a significant amount of the first-year content involves mathematics. Most universities will specify an A level (or equivalent) Mathematics grade in their entrance requirements, so if you are interested in becoming an engineer, you will need to study mathematics. If you want to be involved in production or design but do not want to study mathematics, you could look at alternative courses such as product design.

Very few universities require students to have an A level in Further Mathematics or the equivalent in order to be considered for engineering courses because some schools don't offer it as an option. If they do not, they will generally be teaching you the equivalent during your first year of study. If you have an A level in Further Mathematics, and it is not a prerequisite, you are neither in an advantaged nor disadvantaged position in terms of application, though you are in an advantaged position once the course has started.

> 'Mathematics is essential and takes time to learn. A student who is competent in mathematics can probably learn any other engineering subject comparatively quickly. Frequently, when students ask partners from the engineering industry how much of the mathematics that they studied at university they actually use in practice, they are told "about 10%". This is a most misleading response. Mathematics trains students and graduates to think logically in everything that they do, including writing technical reports. This is one of the reasons why engineering graduates are sought after by so many different sectors of industry. So, although they may not use Laplace Transforms every day, the fact that they studied them in depth once will have developed their methodical approach to their work that avoids them making mistakes.
>
> 'Most universities provide additional mathematics support to engineering students. At the University of East Anglia (UEA) our Learning Enhancement Team employs experienced maths tutors as part of the learning support team, and they in turn co-ordinate a group of senior students who also offer help and guidance. So, a student who is prepared to work hard to develop their mathematics skills will usually find there is plenty of help available.'
>
> *Lawrence Coates, Professor of Engineering,
> University of East Anglia*

Methods of assessment and study

Most universities award degrees on the basis of examinations that are sat throughout the course, although the weightings between examinations sat in the different years of the course may vary from university to university. Most engineering courses also involve coursework, dissertations or practical assessments. Details are given on the university websites. Courses will also contain different amounts of laboratory or practical work, work placements with engineering companies, or on-site work experience. This will also be detailed in the course outlines on the university websites, so you can make sure that your chosen course suits your own preferences or requirements before applying. Teaching is normally conducted through lectures (sometimes supplemented by one-to-one tutorials) or laboratory workshops.

Some courses also have a work placement year as part of the course. For example, the University of Sussex has either a four- or five-year Mechanical Engineering (with an industrial placement year) BEng/MEng that, in Year 3, offers students the opportunity for a paid Year in Industry, which has been shown to help progress a graduate's career at the end of their course. Recent undergraduates have gone on to work for EDF Energy and GE Aviation. Other opportunities at other universities are available, and some even count the placement year as part of your assessment.

BEng v. MEng courses

BEng courses are generally three years in length, and the outcome is a bachelor's degree in the relevant area of engineering. MEng courses are generally four years in length and integrate the bachelor's degree with a master's degree. This is discussed in more detail in Chapter 10. Students who study MEng courses have a more direct route to Chartered Engineer status (see page 130) and will usually command a higher starting salary, as well as gain transferable skills such as teamwork and problem-solving, because the master's component involves more group and project work. In most cases, students opting for the three-year BEng course have the opportunity to transfer on to the MEng course after Year 2.

Combined honours courses

There are a number of joint or combined honours degrees available, but there is less flexibility with engineering courses than there might be with arts or humanities subjects. This is simply because engineering degrees tend to lead towards careers in engineering, and so they focus on this.

Many engineers end up running their own engineering company or business or taking a managerial role within an engineering firm, and so a wide range of degrees that combine engineering with management are available.

As an example, take the Electrical and Electronic Engineering with Management course offered by Imperial College. The programme for the first two years concentrates on the theoretical and technical aspects of electrical and electronic engineering. In the third and fourth years (there are four years because this is an MEng, not a BEng course), there is more flexibility for you to steer the degree towards your own interests, and you would study business courses alongside the engineering courses.

1| Studying Engineering

If you intend to apply for a joint or combined honours course, you must ensure that your personal statement addresses both aspects of the course (see Chapter 5).

Other examples of combined honours courses (and this list is not exhaustive):

- Civil and Architectural Engineering – University of Bath
- Computer and Internet Engineering – University of Surrey
- Electronic Engineering and Computer Science – Aston University
- Energy Engineering with Environmental Management – University of East Anglia
- Engineering and Design – University of Kent
- Engineering and Physical Sciences – University of Birmingham
- Mechanical Engineering and Aerospace Engineering – University of Southampton
- Mechanical Engineering and Languages – Heriot-Watt University

Other courses

Foundation degrees

Foundation degrees are two-year, full-time (or three-year, part-time or distance learning) courses that are provided by some universities. (Do not confuse these with the foundation courses offered to some international students in place of A levels or the equivalent.) They are intended for students who do not have conventional academic backgrounds, for example, students who left school after taking their GCSEs and have been working in a relevant field or mature applicants. Many employers will accept Foundation degrees as an acceptable qualification, and there are many opportunities for students with a Foundation degree to follow this with an extra year of university study to gain a bachelor's degree. You can apply for Foundation degrees through UCAS (www.ucas.com). Subjects available at the Foundation degree level include all the major engineering fields.

Degree Apprenticeships

The vocational higher education market is growing in the UK, with more emphasis placed on teaching practical skills alongside qualifications, with a view to addressing the high-level skills gap that exists in certain industries. Degree Apprenticeships are the older sibling of the T level and combine full-time paid employment with part-time study, enabling students to have the benefit of experiencing both work and study. This ideally suits anyone who is more practically minded and ultimately wants to start in the job market. The programmes will take from three to six years depending on the type of course you choose.

Engineering is typically an industry that lends itself well to Degree Apprenticeships, and there are currently around 1,000 available across the spectrum of engineering qualifications, and the government aims to grow this number. There are currently over 100 universities implementing Degree Apprenticeships in the UK, and the number of employers signing up to the scheme is growing. Roles vary from Materials Process Engineer to Rail and Rail Systems Engineer.

To see what options are available, visit www.ucas.com/apprenticeships /degree-apprenticeships or www.instituteforapprenticeships.org/ apprenticeship-standards/?routes=Engineering-and-manufacturing &levelFrom=5&includeApprovedForDelivery=true.

Some examples of Degree Apprenticeships are:

- Aerospace Engineering – BAE Systems
- STEM Apprenticeships – E-ON Energy Solutions
- Project Support – EDF Energy
- Engineering Degree Apprenticeship Programme – Rolls-Royce

How much will a Degree Apprenticeship cost? Nothing. Employers are expected to cover the cost of apprenticeship training, and you will even be paid a wage while undertaking the apprenticeship scheme. This will, however, mean more competition for places. NB: You are not guaranteed a job at the end of a Degree Apprenticeship, but you will be employable within a skills-based sector, which will give you a significant advantage. Consult www.prospects.ac.uk/jobs-and-work-experience/ apprenticeships/degree-apprenticeships for more information.

Higher National Diplomas

Higher National Diplomas (HNDs) are usually two-year courses, often equivalent to, or taught simultaneously with, the first two years of a bachelor's degree. Students who are successful on the HND course can study for a third year to gain a bachelor's degree. Entrance requirements are normally less stringent than for a degree. For example, Coventry University asks for BBB (120 UCAS Tariff points) from three A levels (including A level Mathematics and/or Physics) for entrance onto some degree-level courses but only 40 points from two A levels for the equivalent HND course.

Scottish degree courses

Undergraduate degree courses at Scottish universities leading to bachelor's degree qualifications are four years in duration, although it is sometimes possible to enter in Year 2. This is sometimes called 'Advanced Entry', and the advantage of this is that it reduces the course length to three years for a BEng course or four years for the MEng course. The Scottish universities that offer this option will specify the entry requirements for the second-year (level 2) entry on their websites. The grade or score requirements are generally higher than for the first-year (level 1) entry. For example, the University of Strathclyde asks for ABB–BBB at A level for first-year entry onto the Electronic and Electrical Engineering course and AAA–ABB for entry into the second year.

The course structure of the four-year degree allows students to study a broader range of subjects in the first year, compared with what is on offer in the three-year degrees common in England, Wales and Northern Ireland. The MEng qualification normally takes five years in Scottish universities.

Case study

'I suppose I had decided on an engineering degree when I was in my early teens. At home, I was always practical and took things to pieces. I made structures out of Meccano and Lego, and learnt many wood-working skills using a variety of aged hand tools. I brewed homemade beer and wine, and developed a keen interest in viticulture. In the garden, I would build walls and concrete paths. I had an old computer and would take it to pieces and then rebuild it in a different way – so emerged my love of all things engineering, from civil to chemical, and from mechanical to electronic!

'At A level I studied Double Maths, Physics and Chemistry (although I found chemistry extremely challenging), and so my pathway to an engineering degree began. My memory of the first term was a huge emphasis on maths. Every day we had a maths lecture interspersed by sessions on Industrial Relations and some project and lab work. My strong advice to those going for an engineering degree is to ensure the university course includes lots of practical work, as this is by far the best way to learn engineering in my view.

'I was fortunate to then join the Army, and during that time had the opportunity to further my studies on a Masters course in Electronics and Guided Weapons. This further renewed my great love of electronics, as I studied Parallel Processing for Robotic Systems, something I then used when I joined the teaching profession to teach Systems and Control.

'Would I change anything? Not a chance! Life as an engineer has been incredibly fulfilling and has enabled me in later life to tackle many things, from car maintenance to house restoration, with confidence.'

John Southworth, MSc in Electronics and Guided Weapons

2 | Building blocks: Getting work experience

Work experience that is related in some way to engineering is an invaluable way of demonstrating to the universities that you are committed to the course. It also shows that you have researched how engineers translate their academic studies into the practical skills that are required in the real world. Work experience will also show you whether you are suitable for a career in engineering and what qualities are needed in a successful engineer. One of the things that an admissions tutor will look for is how serious you are about your chosen course. By writing about your work experience and how what you saw relates to what you enjoy studying or to things that you have read about, you can show the selectors that you have thought about engineering as a whole and as a potential career, rather than as an academic discipline only.

Many surveys have highlighted the importance that engineering employers attach to internships and work experience. And this is also true of your university application. As more and more students chase a fixed number of places, preference is given to those candidates who can demonstrate that they have made an effort to find out what working within the field of engineering will be like.

Engineering affects every aspect of our lives, and engineers make things that need to be commercially viable. Therefore, gaining work experience that is in some way related to engineering should not be difficult. If you are lucky enough to have some connection with an engineering firm, then use this as a starting point. But there are many other part-time jobs and work placements that will allow you to investigate or experience aspects of engineering, such as:

- any company that physically makes something that is then sold commercially – light bulbs, toys, jam jars, window frames, car accessories;
- farm work;
- shops that sell or repair computers or mobile phones;
- a local architecture practice;
- a local builder.

What will you gain from work experience?

- You will have lots to write about in your UCAS personal statement, and you will be able to demonstrate your research into engineering.
- It will help you to decide which area of engineering is most suitable for you. Do you want to work outside or in an office? On practical problems or on the theoretical side of the subject? In a small business or a large multinational company?
- The contacts that you make during your work experience may be helpful in your future career.
- You will find out whether you would actually *enjoy* working as an engineer. Be aware of the quote from the American inventor and engineer Thomas Edison: 'I never did a day's work in my life. It was all fun.' If you don't think being an engineer would be fun, then it is probably not the right career for you.

Looking for work experience

Where to start

You may be fortunate in that your school will arrange this for you as part of a work-experience scheme. If not, then you will need to look for placements yourself.

So, how do you get work experience?

- You could approach local companies or use any contacts that your family may have.
- The institutes of engineering have contacts with engineering companies, so try to go through them (see contact details in Chapter 11). Some will also offer summer courses or 'taster' programmes. The Year in Industry programme provides help and advice for students aiming to use their gap year to gain work experience (www.etrust.org.uk/industry-experience). InvestIN also runs specific work experience programmes (www.investin.org).
- There are many schemes operated by universities and the engineering institutes aimed at attracting students into engineering, and these often involve work placements.

If you know someone, or of someone, in a local engineering firm, try asking to go in for one or two weeks' work experience or work-shadowing during the holidays. Remember that even a single day of work-shadowing is better than no evidence of experience within the workplace, and something even tangentially related to engineering is better than no work experience at all. Helping organise the files for a local car mechanic would give you some access to the practicalities of engineering problems and would be a good stimulus for reading more about automotive engineering, for example.

Online work experience

Through the pandemic, work experience was largely cancelled. Nevertheless, while virtual work experience does not always lend itself well to engineering, it works very well for some companies, with civil engineering firms offering virtual training programmes with modules around Computer Artificial Design (CAD) programmes.

NB: Try to avoid virtual work experience that has a fee attached – there is no need to pay extortionate sums to get what you can achieve yourself.

For information on the programmes available, look at www.SpringPod.co.uk, which gives links to a number of virtual courses; for example, the RAF offers a virtual experience for engineers.

Students should also focus on making the most of YouTube to watch engineering videos. Also, it's well worth checking out Imperial College London's Public Lecture series (www.imperial.ac.uk/computing/outreach/public-lecture-series).

Summer and 'taster' courses

Another good way to learn more about engineering is to use part of your summer holiday on an organised programme that gives you a taste of what a career in engineering entails. A Headstart (www.etrust.org.uk) course is one of many options available and is highly recommended by university admissions tutors. The UCAS website's taster course search page is a good starting point: https://www.ucas.com/undergraduate/what-and-where-study/subject-tasters.

How to apply for work experience

You need to prepare a curriculum vitae (CV), sometimes called a résumé. This should be short and to the point, outlining your education, experience, achievements and contact details. It should be word processed on plain paper (but can be accompanied by a handwritten letter), and it should not be longer than two sides.

2 | Getting Work Experience

Your CV should include:

- full name, address, telephone number and email address;
- synopsis – two lines telling an employer what you want them to know about you;
- education – places, qualifications and grades (start with the most recent);
- skills (e.g. computer skills, software packages you are familiar with, languages spoken, whether you hold a driving licence);
- work experience (full-time or part-time, with names and addresses of the companies or businesses and a brief description of your responsibilities);
- positions of responsibility;
- hobbies;
- names and contact details of two or three people who can act as your referees.

Points to remember:

- highlight any experiences or achievements that are relevant to engineering;
- highlight any skills gained (teamwork, communication, responsibilities);
- ensure that the layout makes it easy to read;
- check all spelling and grammar carefully;
- make sure there are no gaps in the CV (periods of time that are unaccounted for).

A sample CV

Lay out your CV clearly and logically, avoiding gaps and including any exams you are studying for as well as those taken. Below is an example.

George Edward Burrows

Address: Sharow Hall, Aylesford, Kent, ME20 MJB
Telephone: 0123 456 7890
Email: robinedward@gmail.com

Bright and enthusiastic student with a strong skill set in scientific research and innovation. Proven track record of people skills and teamwork.

EDUCATION

2018–25: Wellington College
2025: A levels to be taken: Mathematics, Computer Science, Physics
2023: GCSEs: Mathematics (9), English (6), Geography (8), Dual Award Science (8, 7), Electronics (7), Spanish (7), Psychology (6)

WORK EXPERIENCE

Summer 2024
Mercedes Benz Service Centre
I repaired and serviced cars. I learnt new procedures: how to top-up AC systems with Freon, how to adjust ignition timing and how to take apart, reassemble and change parts in both manual and automatic transmissions.
Summer 2023
Aston Martin Service Centre
I memorised the different models, the variations between them and the details of spare parts. I learnt wheel alignment, how to apply diagnostics and to balance wheels with carbon ceramic brakes, as well as how to buff a car's paint-job. Most crucially, I learnt the team-based aspect of this job, understanding that it is the sum of all parts that makes something work; I was commended for the role I played.

SKILLS

Modern languages: Spanish (intermediate)
IT: literate in all Microsoft programmes and capable of coding in C++, Python and Java

POSITIONS OF RESPONSIBILITY

Captain of 1st XV rugby team; prefect.

INTERESTS

Sport, travelling and understanding other cultures and peoples, innovation.

REFERENCES

Available on request.

The covering letter

The covering letter should add more detail to the CV and also explain why the post you are applying for is suitable for you. You should find out the name of the person who will read it and refer to them by name. If you use the person's name, sign off with 'Yours sincerely'. If for some reason you need to address it to an unknown person, then use 'Dear Sir' or 'Dear Madam' and end it with 'Yours faithfully'. If you are asked to email your application, send the letter and CV as attachments and give a brief overview (name, why you are looking for work experience, brief details of current studies) in the body of the email.

A sample covering letter is shown below.

Mr P Cookson
Signet and Swan Automotive Engineers
Exeter
Devon

1 March 2025

Dear Mr Cookson

I am interested in applying for a temporary role or internship at Signet and Swan Automotive Engineers to gain some experience before I apply to read Mechanical Engineering at university. I am especially interested in Signet and Swan Automotive Engineers, as I admire a lot of the advancements you have made to high-performance cars in the past three decades, not least the development of the Caterham brand. This type of motorsport engineering has formed the basis of my EPQ, in which I am evaluating the future development of Formula E balanced against the loss of classic cars.

In addition to my qualifications shown on the attached CV, I have been teaching maths part-time to schoolchildren, through which I have learnt the importance of taking responsibility. At school, I am part of the Engineering Society, which invites speakers to talk to us once a month. Incidentally, Alfred Robin, one of your principal partners, was at the Goodwood Festival of Speed this year, and I had a good chat with him there about potential opportunities and advancements throughout the sector, and he agreed to do a talk in our school.

I hope that you will be able to consider me for a post. I can be contacted by telephone or email at your earliest convenience.

Yours sincerely
Baxter Fredericks

Work experience interviews

If you are lucky, you may then be called for an interview. Most of the advice given in Chapter 6, on university interviews, is also relevant to job interviews. Other points to remember are listed below.

- Do some research on the company in advance. Learn about what they produce and how long they have been operating.
- Ensure that you can explain how your particular skills and qualities would be useful to them.
- Be clear about when you can start, how long you can work for and whether there are particular periods when you cannot work, for example, on the day your examination results are released.
- Wear smart clothes, a business suit if possible. Ensure that you have polished your shoes and don't wear too much jewellery.
- Introduce yourself, address the main interviewer by name and offer your hand for a handshake at the start and again at the end.

How to use your work experience effectively

Undertaking work experience is a means to an end – to show the university selectors that you are a serious candidate. You will be expected to write about what you discovered about being an engineer in your personal statement and to discuss it in more detail at your interview. It is useful to keep a diary of your experiences while you are there. Record what you did, saw and heard and, in particular, conversations you had with engineers – not just technical things but also about their training and what they see as the main challenges and rewards in working in an engineering field. It is also a good idea to then do some research or reading on things that you encountered during your work placement.

A common mistake that people make when trying to design something completely foolproof is to underestimate the ingenuity of complete fools.

Douglas Adams, author

Case study

Why I chose engineering

'I chose engineering as it was already deeply linked with my childhood interests and my family history. I was always interested in lego from a young age, and was interested in more challenging constructions; I would construct lego buildings intended for individuals aged 16+ when I was around 8/9 years. My favourite part about lego was to follow the instructions in order, keep the lego blocks organised in size or colour and finally trust the process.

'All of my ancestors were from the mountains, from the village of Khaibakh in the high Chechen mountains. Before the massacre, it flourished as a safe and extremely beautiful settlement for over 40 families. In the middle of the village was a tall tower, made out of randomly sized mountain rocks; this tower was known to be built by my ancestors and played the role of the safe house of the village. This inspired me to further study the crafts and skills.'

My A level subjects

'I studied maths and physics to help me build a solid foundation for my future understanding of engineering theory. I also did business to help me with the project management side of the industry. I also did art as one of my 10 GCSE subjects, which I believe will help me with the visual and architectural side of engineering.'

How work experience helped

'Work experience in the civil and structural engineering field, alongside working in a project management company, has given me a clear idea of what the industry expects of an engineer, what the working environment was like and all the safety regulations that have to be strictly followed. This gave me the certainty I needed to be 100% sure that's what I aim to be in the future.'

Why I made my course choices

'I chose three civil engineering courses; however, I also applied for one course in general engineering as this would allow me to explore a wider range of engineering careers. For example, during my work experience, one of the civil engineers shared that he did a general engineering degree that he was able to adapt to a civil working environment. Finally, I also chose a general engineering and architectural design course, which would allow me to explore the architectural side of work. This course would allow me to combine two of my greatest interests and help me develop the key skills necessary without having to prioritise one course over the other.'

Bisni Khaibakh, UCL General Engineering and Architecture

3 | Measure twice, cut once: Choosing your course

Choosing universities and engineering courses can be a bewildering experience because you will be confronted by an enormous number of options. This chapter covers the steps that you should take to narrow your search and to find the most appropriate courses for you, including what to look for in the course content of an engineering course, academic requirements, types of universities and location. It also covers what the course or university can offer you in terms of your future career and what you will gain from it on a personal level.

What to consider

Academic considerations

You are allowed five choices on the UCAS application. The basic factors to consider when choosing your degree course are:

- the engineering course you are looking for;
- where you want to study;
- the kind of university you are aiming for;
- your academic ability.

Non-academic considerations to bear in mind include:

- accommodation;
- location;
- sporting, musical or other extracurricular facilities;
- costs.

You need to think seriously about your choice of universities, as the decisions you will take now may determine your future career options. At this stage, you may already have an idea of which universities you want to consider, based on the advice of friends and/or family, but you need to be as open-minded as possible. Make a list of between 10 and 20 universities in which you are interested; it is then important to reduce this to a much shorter list. Not only will the university be where you begin the next step in your education, it will also be your home for three or four years, so think carefully about the location, environment and accommodation options, as well as the suitability of the course(s) on offer.

Here are some things to research and consider for each university.

- Get hold of the prospectuses and any departmental brochures for more details. Remember that university publications are there to attract applicants as well as to provide information and may be selective about the information they provide, so read all of it bearing this in mind.
- Visit the websites of the universities you are considering. This is the best place to look for the most current information about a university. Another useful element of university websites is information on past and present students from a range of disciplines who give their views on student life at the institution. Some university websites have email links to current students who can answer any questions that you may have.
- Find out when the open days are and go to them if you possibly can. You will have a chance to look at the facilities, talk to current students and find out more about the course. Virtual open days are commonplace nowadays, with many universities offering this opportunity alongside their traditional open day event. Check university websites for when these are happening, as they provide valuable insight into university life.
- Discuss engineering and studying it as a subject with engineers you know; ask for their views on the reputations of different universities and courses. This may bring up some highly rated engineering university departments you may not have thought of.
- Investigate the grade requirements and be realistic about the grades you are expecting – your teachers at school or college will be able to advise you on this.
- Check that the course allows you to choose the particular options in which you are interested. If you are considering, perhaps, mechanical engineering but have a particular interest in aeronautics or automotive engineering, make sure that these options are available. You will not always know what each option actually covers by its title, so read the department's own prospectus carefully and address any unanswered questions by contacting the admissions tutors directly – contact details are usually on the departmental website.

- Think about whether you would like your course to include a placement with an engineering company. This might be for a few weeks, a term or even a year. If this is something that interests you, find out who organises the placement – you or the university – and whether there is any funding to cover, for example, travelling costs or subsistence.

Consider whether you want to spend some time abroad. Many engineering courses offer the option of a year abroad, studying at a partner university. If so, think about practical details such as language or visa requirements and costs.

- Investigate how much practical/laboratory work is included in the course and what the practical engineering facilities are, for example, whether the facilities include state-of-the-art machinery or testing equipment.
- Look at what IT facilities are offered. If you do not have your own laptop, will the university have facilities for you to manage without one of your own? Access to computing facilities can be very important when you are working on a dissertation. Find out whether there is Wi-Fi access in the study areas and/or accommodation.
- Ask about the reading lists and whether the books are available in the library. Are you required to buy your own textbooks? If so, are second-hand copies available?
- Look up the specialities of the engineering course staff. Are they experts in the field of engineering in which you are particularly interested? Use the internet to find out what their experience is and what they have published, as this will give you a better indication of a department's strengths.

Student experience

'I needed a lot of university guidance to understand where the best place was for me to study at university. There are so many fantastic engineering degrees. My best advice to anyone is to, if possible, visit the campuses and ignore league tables – choosing a university is so much more than just the course, it is about whether the university offers the right experience for you. And I did not get it wrong.'

Joey Relisan, University of Southampton,
BEng Electronic and Electrical Engineering

League tables

Newspapers often feature university rankings or league tables, but there is no official ranking of universities or university courses in the UK, so these tables are created using criteria selected by the newspapers themselves. There is a significant amount of variation between these tables because each table will score the universities in a different way.

However, as long as you approach these rankings with caution, they can be a useful aid to the selection process, particularly if you look at how the rankings are assessed rather than simply looking at a university's position in the tables. There will be some criteria that you might regard as being important to you – graduate job prospects, for example – while you might not be so interested in the student-to-teacher ratios.

A good starting point is the *Guardian* newspaper's university rankings. The general rankings can be found at: www.theguardian.com/education, and there is a link to the subject-specific rankings, which feature a number of engineering disciplines. Of particular interest might be the scores based on job prospects and the student satisfaction with the course and with the teaching.

You might be interested in how the UK's universities are regarded on the global stage, particularly if you are an international student or you are planning on working abroad at some stage in your career. There are a number of world university rankings that you might find useful. But bear in mind that, like the UK university rankings, they are not in any way 'official' rankings and are based on criteria that the organisations compiling them see as being important.

One example is the world university rankings compiled by *Times Higher Education* (www.timeshighereducation.com).

There are many other league tables and rankings, for example, the one produced by *The Sunday Times* newspaper for UK universities or by QS, which specialises in business education, for world rankings. No two rankings will produce the same results, so you need to use them as part of the process in making your decisions, not as the sole reason for a choice of university.

Another useful reference point is the National Student Survey, which offers a student perspective on individual degree courses at any given university. The questionnaire is completed by graduates across the UK; the results are published annually and can be viewed online at www.officeforstudents.org.uk/data-and-analysis/national-student-survey-data/

After completing your research, you should be able to narrow down your original list to the five choices for your UCAS application. Once you have done this, discuss the list with your teachers to see whether they think it includes sensible choices. They may ask you to think again about some of the choices.

Common areas of concern for a school's UCAS adviser are:

- your current subjects don't fit the requirements of the courses you are applying for;
- that the grade requirements are too high (or too low) for the applicant's likely academic achievements;
- that there is too much variation within the choice of courses to enable the applicant to write a coherent and focused personal statement (see Chapter 5).

Choosing the right course

Most universities offer a wide range of engineering courses, and it is important to investigate these thoroughly before making a choice. A particular university might list the following on the UCAS website:

- Aeronautical Engineering
- Aeronautical Engineering with a year abroad
- Biomedical Engineering
- Biomedical Engineering with a year abroad
- Chemical Engineering
- Chemical Engineering with a year abroad
- Chemical with Nuclear Engineering
- Civil Engineering
- Civil Engineering with a year abroad
- Electrical and Electronic Engineering
- Electrical and Electronic Engineering with a year abroad
- Electrical and Electronic Engineering with Management
- Information Systems Engineering
- Materials Science and Engineering
- Mechanical Engineering
- Mechanical Engineering with a year abroad
- Structural Engineering

University websites and prospectuses can provide more details on specialisms, and it is well worth spending time going through these. Some universities offer integrated or general engineering courses that allow students to choose their specialisation later in the course. If you are truly undecided about which particular area of engineering is the most suitable for you, then this might be a good option. However, even if you apply for an integrated or general engineering course, it would be a good idea to have an idea of which specialisations interest you most, as this will probably be asked at the interview.

3| Choosing Your Course

When considering possible courses, read through all of the course content. Do not choose a course just because of its title. Courses with the same name at different universities can vary immensely in their content, and within the courses themselves the likelihood is that you will have a range of options to choose from once you start your course. This is also important if you are interviewed (see Chapter 6) because you may be asked to justify your choice. Being able to discuss the course structure in detail will be an important factor in convincing the interviewer that you are a serious applicant.

Similar-sounding courses also do not always have the same entrance requirements (examination results and preferred A level subjects). Examination results are specified as either grade requirements (e.g. AAB) or Tariff points (e.g. 136 – see the section on the UCAS Tariff below). Unless you are applying post-results (as a mature applicant or during your gap year), your referee will be asked to predict the grades that you are expected to achieve in your examinations. You should find out in advance what he or she is going to predict, because this will determine your choice of universities and courses. For example, if you apply for five university courses that require AAB but your A level predictions are BBB, there will be a high chance of being rejected by all of your choices. You will then have to try to find alternatives through the UCAS Extra scheme or through Clearing (see Chapter 8). Similarly, if you are predicted to achieve A*AA, you are probably aiming too low if all of the courses you are applying for require CCC at A level.

As a rough guide, if you are predicted, say, ABB, it would be risky to choose a course that requires AAB. It would be safer to choose three or four that require ABB and one or two that require BBB. This not only means that you have a good chance of getting a number of offers, but it also gives you options if you do not quite meet the grade requirements (see Chapter 8).

The UCAS Tariff

While some universities make offers based on grades at A level, Scottish Highers or IB scores, others make offers using the UCAS Tariff. The UCAS Tariff is a points-based system that compares different education systems or achievements. For example, ABB at A level is equivalent to 128 Tariff points, roughly equivalent to an IB Diploma score of 27.

UCAS Tariff for A levels

A* grade – 56 points
A grade – 48 points
B grade – 40 points
C grade – 32 points
D grade – 24 points
E grade – 16 points

> AS scores are now worth 40% (or as close as possible) of the full A level UCAS Tariff points at each grade. D* in the BTEC and H7 in IB is valued at 56 points, and a top grade in a Scottish Higher is 33 points. An EPQ is 50% of an A level, and an A* in the EPQ is worth 28 points.
>
> **UCAS Tariff for T levels**
>
> Distinction – 144 points
> Merit – 120 points
> Pass (C or above on the core) – 96 points
> Pass (D or E on the core) – 72 points
>
> Full details can be found on the UCAS website at: www.ucas.com/ucas/tariff-calculator.

Sandwich courses

Sandwich courses, either BEng or MEng, offer students the chance to spend some time gaining industrial experience through work placements organised or suggested by the university. Typically, this might involve taking a year away from the university, although some universities will also offer a variety of shorter placements. Students who choose sandwich courses do so because:

- they want to gain experience of working to enhance their future job prospects;
- they want to learn more about what working as an engineer entails;
- they want to make contacts in the engineering world to help them to find jobs in the future.

Overseas study

Studying abroad could also be a factor that affects your degree selection. It is possible to study engineering in many countries as part of a degree based at a British university. Not all of these courses send you off for a full year, though there are schemes that last for only one term or semester. You do not need to be a linguist either, as it is always possible to study overseas in an English-speaking location such as North America, South Africa, Australia or Malaysia. As an example, if you study Electrical and Mechanical Engineering with International Study at the University of Strathclyde, you will have the opportunity to study at a partner university in Year 4. The university's partner universities include institutions in Canada, the US, China, Singapore, New Zealand and Australia.

The availability of student exchanges increased through programmes such as Erasmus+, which encouraged universities to provide international opportunities where practical – particularly in Europe.

The popularity of overseas study has encouraged some universities to develop special exchange relationships with universities further afield.

As a result of Britain leaving the European Union in January 2021, the Erasmus+ programme has been replaced by the government's Turing scheme. The UK government claims it 'remains committed to international education exchanges, both with the EU and beyond'. For further information, please see www.gov.uk/government/publications/turing-scheme-international-study-and-work-placements

Many universities will provide contact details or testimonials from students who have studied abroad as part of their degrees. For example: www.strath.ac.uk/studywithus/studyabroad.

Academic and career-related factors

Academic ability

It is important to be honest about what you think you will achieve in your A levels or equivalent because, for most, this is the deciding factor for selection. The best way to get a strong sense of your predicted results is to speak to your teachers.

Remember, be realistic; you may think that you can do much better than your teachers' predictions or than your AS (either internally or externally assessed) results indicate, but the predicted grades and previous results will go on your UCAS application form, and so it is important to ensure that the courses you apply for are consistent with your likely results. Check also whether the universities are likely to make a grade or Tariff points offer. If the grade requirement is, for example, AAB, don't assume that a combination of grades that gives you the same Tariff points (136, achieved by gaining A*BB) would satisfy the university's requirements.

For more details about UCAS and filling in your application, see *How to Complete Your UCAS Application: 2026 Entry* (Trotman Education).

Educational facilities

Take a look at the facilities the university has to offer. Here is a checklist of what to look out for:

- access to lecturers if you need help;
- computer facilities;
- course materials;
- laboratory/workshop provision;
- lecture theatres;

- library facilities;
- multimedia facilities;
- study facilities.

Quality of teaching

The Office for Students, the Higher Education Funding Council for Wales, the Scottish Funding Council and the Department for the Economy assess the level of teaching across the UK. Their findings are publicly available – see www.officeforstudents.org.uk, www.hefcw.ac.uk, www.sfc.ac.uk and www.economy-ni.gov.uk. League tables normally incorporate these into their rankings. In September 2017, the government introduced the Teaching Excellence Framework (TEF), which is there to assess universities and colleges on the quality of their teaching and is intended to help students with their university choices. It is worth noting this is a voluntary scheme, and universities can receive a bronze, silver or gold award reflecting their excellence in teaching, learning environment and student outcomes. Without a TEF award, universities cannot charge a student more than £9,000 per year.

Type of institution

There are four types of institutions from which you can obtain a degree:

- 'old' universities;
- 'new' universities;
- private universities;
- higher education colleges.

The 'old' universities

These are traditionally seen as the most ancient universities (such as Oxford, Cambridge, Edinburgh, University College London, Imperial College London and Durham), along with those established in the early and mid-20th century (such as Bristol, Leeds, Liverpool and Reading), plus those established in the 1960s (such as Sussex, Bath and Stirling); usually with higher admission requirements and with a strong emphasis on research.

The 'new' universities

Pre-1992, these were polytechnics, institutes or colleges, for example, Kingston, Central Lancashire and Westminster. These tended to focus more on vocational courses with less emphasis (or none at all) on research. There are a number of excellent engineering degree courses at new universities that are very well regarded and highly competitive to get into, and they are often more flexible in terms of sandwich courses, work placements or links with industry.

The private universities

New private universities continue to join the university landscape and can offer well-funded departments, though with a high price tag. At this stage, there are no private universities offering undergraduate bachelor's degrees in engineering, though Bucks New University offers Engineering Design and Arden University offers an MSc in Engineering Management. This may change over time.

Colleges of higher education

These are specialist institutions that have links with universities. The university awards the degree and can deliver part of the course, along with the institution. Many colleges also offer pathway programmes, such as Access or foundation courses for students who do not have the necessary academic qualifications for direct entry to a university.

Non-academic considerations

The starting point for your research ought to be the type of course you want to follow. Once you have done this, you should then look at the non-academic facets of the universities that you are considering. These are irrelevant to the course but instead concern what the university can offer you on a personal, social and cultural level.

Finances

Finance is an important factor to consider, as you will need to juggle a lot of outgoings when you go to university. You will need to take into account:

- accommodation costs;
- availability of part-time work in the area to earn some extra money;
- living costs, such as food;
- 'lifestyle' costs;
- travel from your accommodation to the university during term time;
- travel from your home to and from the university for holidays;
- proximity to your home, family and friends – will it cost you a lot of money to visit friends or to go home?

Accommodation

The university is likely to be your home for three or four years, so think carefully about the location, environment and accommodation options. Accommodation can vary wildly between institutions, so you will need to think about where you would feel most comfortable. Do you want to live in halls of residence with other students or independently (or with friends) in rented houses or flats? Do you want to be near your lectures or are you happy to live further away from the university?

Universities all have accommodation offices that provide help or information for students who do not want to, or are unable to, live in university-provided accommodation. Most universities offer arranged accommodation for first-year students in halls of residence, which may either be owned by the university or be shared with other institutions, but often students are expected to find their own accommodation in subsequent years.

Entertainment

You will be spending the next few years in a new place, so you will need to have a look at the entertainment facilities it has to offer both within and outside the university. Are your particular interests or hobbies catered for? If you're a keen sports enthusiast, have a look at the facilities on offer and the sorts of clubs and teams you can join.

Site and size

- Campus university outside a town or city?
- Campus within a town or city?
- University buildings at various locations within a town or city?
- Large or small?

While some students have a clear picture of where they want to study, others are fairly geographically mobile, preferring instead to concentrate on choosing the right degree course and see where they end up. But university life is not going to be solely about academic study. It is truly a growing experience – educationally, socially and culturally – and so you need to do your research and think carefully to ensure that you choose the best university and course for you.

What can the university offer you?

This is possibly the most important question for you to consider. Universities need students. Even the most popular and oversubscribed universities are constantly looking to attract the best students. League tables and advice from teachers, employers or friends can help you in choosing the most suitable courses, but if you plan your application carefully, you will have some choices to make. In this chapter, we have looked at a range of selection criteria: location, accommodation, the course structure, facilities and other things. You might also want to investigate whether scholarships are available (see Chapter 9) and also the feedback from students who have previously studied at the university. The *Guardian* newspaper's university rankings incorporate scores based on student satisfaction with the course and with the teaching.

> Science is about knowing; engineering is about doing.
> Henry Petroski, Professor of Civil Engineering

4 | Righty tighty, lefty loosey: Completing your UCAS application

This chapter is designed to help you complete your UCAS application. Further advice on filling in your application is given in *How to Complete Your UCAS Application* (Trotman Education).

The UCAS application

The UCAS application is completed on the UCAS website. UCAS Hub is UCAS's online application system, which you can use to register and apply to your university choices. You can sign into your Hub to check your application's progress at any time to view any interview invitations and offers you receive. There are 14 sections for you to complete on the UCAS application.

1. **Personal information.** Your name, date of birth and gender.
2. **Contact and residency.** Your telephone number, postal and email address, your 'nominated access' and residency details. 'Nominated access' is someone who can act on your behalf if you are unavailable at any time during the application period. You need to read the instructions carefully to avoid making mistakes.
3. **Where you live.** A new section from 2025 entry. Where you've lived for the last three years and for what purpose.
4. **Nationality.** Your nationality, questions about UK student visa, settled or pre-settled status in the UK and your passport details.
5. **Supporting information.** Questions about you and your family living or working in the EU, EEA or Switzerland.
6. **English language skills.** Question on whether English is your first language; if it is not, you can submit your International English Language Testing System (IELTS) or TOEFL English proficiency test numbers.

7. **Student finance.** If you are planning to apply for student finance, you can complete this section to make your finance application easier. (UCAS will share your information to make the two processes more streamlined.)
8. **Diversity and inclusion (UK applicants only).** Equality monitoring information that universities only see after a place is secured.
9. **More about you.** Declaration of any physical and/or mental health condition, long-term illness or learning difference.
10. **Education.** This means your examination results, where you have studied and any examinations to be sat.
11. **Employment.** Details of any full- or part-time paid employment. If you have had gaps in your education because you were in employment, you need to give the details here.
12. **Extra activities (UK applicants only).** If you've participated in any activity to prepare for university.
13. **Personal statement.** See Chapter 5 for more information.
14. **Your choices of university.** You are allowed to choose up to five courses. Pay particular attention to the course codes and university codes, and ensure that all of the required information (where you intend to live, to which campus you are applying etc.) is included.

Once you have completed all of these sections, your referee will add his or her comments about your suitability for your chosen courses, as well as your predicted grades. Your referee is normally someone at your school or college (such as a head of sixth form, teacher or tutor), but for applicants who are not at school, this might be an employer (see Chapter 7).

Suggested timescale

Use the timescale below to help you plan your application. See Figure 3 (on page 50) for deadlines.

Year 12

In Year 12, during your first year of A levels or IB, you should start to think seriously about what type of engineering course you want to follow. Talk to as many people as possible – your teachers, family and careers advisers. Don't just focus on what area of engineering interests you most, but also on whether you want a three-year BEng course or a four-year MEng course, whether you want a campus university or one in the centre of a city, close to home or in another part of the country.

Your next step should be to find out when the universities that interest you have open days and arrange to visit them. After this, you ought to be in a position to make up a shortlist of courses and universities, ready for your application. You can start to order prospectuses from the universities or download the PDF versions from the university websites. Remember to make a note of the grade requirements for your chosen universities.

In August, refine your choices in the light of any AS results or predicted examination results.

Year 13

In September of Year 13, you can submit your UCAS application. The deadline for applications for 2026 entry is 14 January, but I recommend that you apply as early as possible. If you are applying to Oxford or Cambridge universities, your application has to be submitted by 15 October 2025. Some universities will want to interview you, and this can happen from November onwards. If you are required to sit entrance tests, you may be asked to do this in October or November.

You should start to receive decisions from the universities from January onwards, although some universities might get back to you earlier than this. You can keep up to date with the status of your applications in your UCAS Hub account. You will receive one of three possible responses from each university:

- conditional offer;
- unconditional offer;
- rejection.

If you receive a conditional offer, you will be told what you need to achieve in your A levels. This could be in grade terms, for example, AAB (and the university might specify a particular grade in a particular subject – AAB, with an A in mathematics), or in UCAS Tariff points (136 points from three A levels – see Chapter 3).

Unconditional offers can be given to students who have already sat their A levels, such as gap year students applying post-results. In recent years, 'conditional unconditional' offers have been used to force students into making a decision early on their university choices. However, the university's regulator has strongly discouraged these offers with fines amid concerns it was undermining the integrity of the university system.

Rejection means that you have been unsuccessful in your application to that university. If you are unlucky enough to receive rejections from all of your choices or you decide to withdraw from your choices, you can then use the UCAS Extra scheme from February to apply to other universities, or to add choices if you did not use all of your five choices in the initial application. Extra allows you to add one new choice at a time. If you are successful in gaining an offer from your first Extra choice and you accept it, you are committed to accepting the place; if

4| Completing Your UCAS Application

you decline the offer or are rejected, then you can approach another institution through Extra.

Once you have received responses from all five universities (or offers through UCAS Extra), you will be given a deadline by UCAS by which time you have to choose one university as your firm acceptance and a second insurance offer, normally one that requires lower grades. (Once you have accepted your first choice university, you will be sent information about accommodation and fees and other practical information directly from the university.)

You sit your examinations in the summer, and receive your results in July or August:

- A level results are published in the third week of August;
- Scottish Higher results are released in the first week of August;
- IB results come out in the first week of July.

When the exam results are published, UCAS will get in touch and tell you whether your chosen university has confirmed your conditional offer. Do not be too disappointed if you have not got into your chosen institution; just get in touch with your school/college or careers office and wait until Clearing begins in early July, when all remaining places are filled. You will be sent instructions on Clearing automatically, but it is up to you to get hold of the published lists of available places and to contact the universities directly.

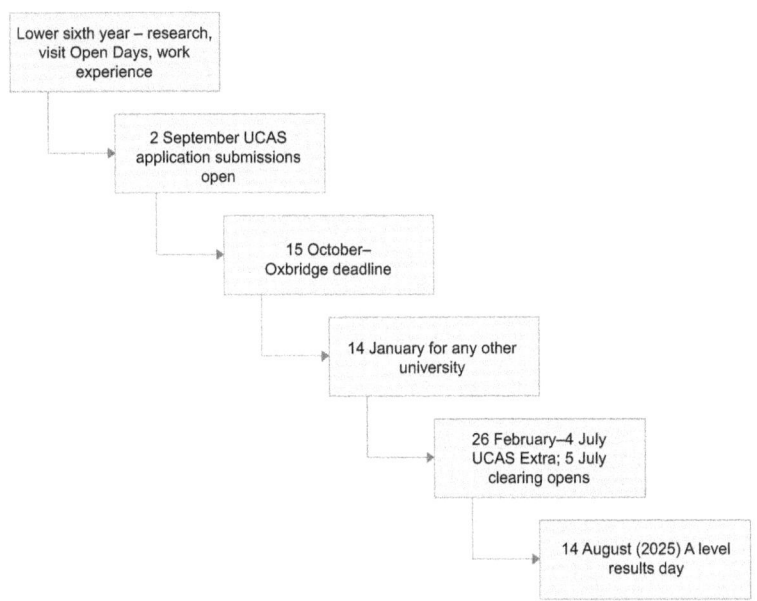

Figure 3 Timeline of applications

Entrance examinations

As competition for places is so fierce, some universities ask applicants to sit entrance examinations as part of the application process.

University of Cambridge

If you are applying for engineering courses at Cambridge, you will be asked to sit the Engineering Specific Admissions Test (ESAT) from November 2024. The ESAT will replace the previously used Engineering Admissions Assessment (ENGAA). The ESAT is a new assessment developed to assess your understanding of key mathematical and physical principles, which are essential for studying engineering. The test is set to include multiple-choice questions on physics and mathematics, designed to assess applicants' problem-solving ability and application of theory in practical contexts.

The ESAT will typically be taken at your school or college. You need to ensure that they register you for the assessment by the deadline in September. The full details of the test format, registration and other specifics will be made available through the University of Cambridge's official undergraduate admissions site closer to the test period.

Additionally, depending on your engineering course choice, you may also be required to sit the Sixth Term Examination Paper (STEP) in mathematics, which tests deeper mathematical knowledge and skills needed for studying engineering at Cambridge.

Students are strongly encouraged to visit the University of Cambridge's website for the latest information and updates about their admissions assessments. You can find more detailed guidance at the official site for undergraduate admissions testing.

University of Oxford

The University of Oxford's Physics Admissions Test (PAT) is a key part of the admissions process for applicants to engineering courses. The PAT is designed to assess both mathematics and physics skills through two sections:

1. **Section A.** This covers mathematics and is specifically geared towards physics. A calculator is allowed.
2. **Section B.** This focuses on physics, and calculators are **not** permitted.

The test lasts two hours in total. However, since 2024, there have been significant changes due to the introduction of an online format. Applicants will complete the test digitally, using Pearson's OnVUE platform. A digital calculator is provided, and physical calculators will no longer be allowed during the test

Students are advised to visit the University of Oxford's website for further details: www.ox.ac.uk/admissions/undergraduate/applying-to-oxford/guide/admissions-tests/pat.

A good resource for sample and practice questions ahead of university interviews can be found at: http://i-want-to-study-engineering.org.

Other universities

It is usual for universities to require you to sit an admissions test for engineering, particularly where A level Mathematics is not present; however, certain universities may interview. Interviews are usually not too formal and are more of a conversation surrounding your reasoning behind choosing an engineering course, as the admissions team wants to ascertain if that is the right route for you. The University of Edinburgh (Global SELECT pathway), Liverpool John Moores University, Birmingham Metropolitan College and some other universities may ask some candidates to sit an aptitude or admissions test (in the course requirements information on the UCAS website, this is listed as 'IOT', which means 'institution's own test'). The tests are only normally required for mature students or students with unusual qualifications, and you will be given information about whether you need to sit the test and its format when you apply. The tests are used as one piece of evidence in assessing candidates, alongside many other criteria such as grades achieved, predicted grades, the personal statement and references. The UCAS website provides a list of courses that may require additional admissions tests: www.ucas.com/applying/applying-university/admissions-tests.

Taking a gap year: Deferred entry

Many students take a gap year between their final school or college examinations and the start of their university course. Universities are nearly always happy with this, as students who take a year away from studies are often more motivated and mature when they start their degree studies. But in terms of the application, a gap year will only enhance your chances of getting a place if you use the year productively.

There are two application routes for students taking a gap year.

- You can apply for deferred entry; that is, you apply in the final year of the A level course for entry a year later. So, if you are sitting A levels in June 2026, you would apply for deferred entry in September/October 2027, not 2026.

- Alternatively, you can apply at the start of the gap year once your A level results are known.

There are advantages to both routes, depending on your plans and A level (or equivalent) grades.

Advantages of deferred entry

- Once you have satisfied your offer, you will know where you will be studying in a year's time, and so you can make firm plans about what you will do during the year.
- You can plan to be overseas (travelling, working, volunteering) without worrying about having to return for university interviews.
- You have a second chance to apply to universities during the gap year if your initial application is unsuccessful.

Advantages of applying during the gap year

- More time to decide which field of engineering really interests you.
- You will already know your results, so you can focus your application on those universities for which you have already achieved the necessary grades or scores.
- If your predicted grades are not high but you feel confident of doing better than your school or college expects, you can avoid the possibility of being rejected on the basis of your predictions rather than actual ability.

'Applicants are led to believe that universities have differing views on the value of a gap year. This is not likely. A student who is taking a gap year to put off starting university and looks set to waste their time will always be frowned upon. But a student who has clear plans of almost any sort will gain much maturity from a gap year. Within a body of Freshers it is always possible to spot the students who have had a gap year. They are more mature and engage in interesting conversations beyond the obvious "What did you get in your A levels?"

'A typical gap year involves working for four to six months to earn money to fund travelling. This can be a good use of time. Some applicants engage in activities for charity which can be equally valuable. Anything that exposes you to new experiences will broaden your mind.

'A common concern with taking a gap year is that the student will forget their mathematics, which is such an important part of engineering, particularly in the first year. Even though it should be like riding a bike, there is no doubt that a year completely away from mathematics does mean that gap year students arrive a bit rusty and often do badly in the self-diagnostic tests that are commonly used at entry. But within a couple of weeks they are back up to speed with no long-term issues. So, this is no reason to avoid a gap year.

'Occasionally there are issues to be aware of associated with accommodation guarantees, etc. So, it is worth keeping in touch with the admissions tutor during the gap year. A significant minority of students complete their gap year and then decide to change degree course. Some universities probably use this as a reason for not encouraging gap years, but on the contrary this is a really good thing. If a student has had time to think about career options and has made a positive decision to change (see Chapter 7) then this is far better than arriving at a degree and realising halfway through the first year (or even later) that it may not be the right one.

'Keeping in touch with the admissions tutor can help both sides. You may learn some useful information that makes the transition to university easier. The admissions tutor is frequently one of your lecturers and will usually welcome the opportunity to get to know some of the students in advance of their arrival. Occasionally they can put you in touch with relevant student societies so that you are already in contact before the mad rush of Freshers' week. A good example is Engineers Without Borders that has branches at many universities and will welcome your attendance at some of their meetings before you arrive.

'Considerable value can be added by keeping a regular log of what you do during your gap year. When completing job applications five years later it can be very helpful to be able to reflect on what you learned from the year; trying to do it solely from memory leads to a less rich account.'

Lawrence Coates, Professor of Engineering,
University of East Anglia

If you are planning on taking a gap year, you must plan it carefully so that you will gain something from it. This could be a life or work experience, maturity, a chance to extend your studies into new areas, independence or money to fund your university studies. A year spent resting and playing computer games, however, attractive that may be after all your hard work at school, is not going to convince the universities that you will be a stronger candidate as a result.

Here is an excerpt from a personal statement about taking a gap year:

I am taking a gap year in order to gain more maturity and experience.

Such a statement is not going to convince the admissions tutors that you have made constructive plans for your gap year, nor is it likely to help you develop or bring new skills and ideas onto their course. A better statement might be:

During my gap year I am going to take part in a voluntary project building small-scale dams with Starfish, a voluntary organisation based in Thailand, in a mountainous area, to help control the water that causes soil erosion and crop damage. We will then work in a small village building toilet facilities for the old people who live there. I am looking forward to being practically involved with these engineering projects and to learn more about problem solving in real situations. To raise money for this trip, I will be working in my local hospital in the maintenance department.

This is much more impressive because the candidate has linked what she will do in her gap year to her future degree course (Civil Engineering), and it is clear that she has thought carefully about what she will do during the year.

A piece of advice: phrases such as 'I have arranged to . . .' are much more convincing than 'I hope to . . .' when discussing your gap year plans.

Student experience

'I never considered taking a gap year during my studies, as I had a clear and definite plan of getting into university after graduating from high school. However, my university applications to the desired universities did not go the way I had hoped, and a deferred entry seemed to be an apt alternative, which presented a unique opportunity of experiencing and trying something new throughout the year. I decided to make the most of this situation and turn it into a worthy and memorable period. The first thing I did was move back to my home country – Kazakhstan – with plans of starting my own business there. A few months were spent on reconnecting with my friends and family, but I also pursued some exciting initiatives. I received a lot of support and inspirational ideas. This has encouraged me not to be afraid of trying. I met some incredible people with brilliant minds. Together we created a few projects to help young people with some innovative learning opportunities. It was good to do something useful for my country, while contributing to my future education and career.'

Alan Kononenko, University College London,
MEng Mechanical Engineering

Gap year plans

It is always a good idea to check with your chosen universities that a gap year is acceptable to them before committing yourself. There is likely to be information on their websites addressing this. If there is none, you can email the admissions staff to ask them. This is particularly important if you are taking a gap year for reasons other than wanting to take a year between school studies and the degree course to gain experience in engineering.

Some of the other reasons for taking a gap year are:

- to work towards extra qualifications because you need to strengthen your application or because you wish to change direction (e.g. if you have studied mathematics up to GCSE only, you will be doing an evening class in A level Mathematics alongside your other projects);
- you started another course (such as a degree course in another subject; see Chapter 7) and then realised that it was not right for you, so you have withdrawn from it;
- you have been working, and you now want to return to studying;
- you may have had an illness or other issues that required you to take a break from studying.

> *Engineers like to solve problems. If there are no problems handily available, they will create their own problems.*
>
> Scott Adams, cartoonist

5 | Fire on all cylinders: The personal statement

Arguably, the most important part of your UCAS application is the personal statement. It is also the one part of the form where you have complete freedom to decide how you wish to demonstrate your suitability for the course to the selectors. You have 4,000 characters to convince your five chosen universities that:

- you have good reasons for studying engineering;
- you have researched your future career thoroughly;
- you have appropriate personal and academic qualities to become a successful engineer;
- you will be able to contribute something to the department and to the university.

Who will read your personal statement?

Before you can write a personal statement, you have to think carefully about your choice of courses. This is because each admissions tutor will read the personal statement with his or her own course in mind, and he or she will expect what you write to be consistent with the course. For this reason, make sure that the five courses you choose are closely linked in terms of course content and outcome. If you are applying for a civil engineering course, the selector will expect to read about your interest in civil engineering, books related to the subject that you have read and relevant work experience. Similarly, an admissions tutor for electronic engineering will expect the personal statement to address this subject. Clearly, you cannot convince both that you are serious about their courses in one personal statement.

While there is a certain degree of crossover between some engineering disciplines, largely they are separate; therefore, it is better to be sure of your choice. If you were combining any engineering disciplines in your application, the link would need to be strong in order to be able to reflect it in your personal statement, such as mechanical engineering and automotive engineering or civil engineering and construction engineering.

> 'Your personal statement is your chance to tell me what makes you special. I can already see your academic profile (past and predicted), so this is your opportunity to tell me how you're going to contribute to the Cardiff academic and social community. Is there something that makes you unique and you think we might value? Tell me about it and how you think this is relevant to studying engineering. Lots of students talk about how well rounded they are, and in the eyes of many admissions tutors, this is important. But once in a while I would like to read about an applicant who is truly special and unique. Have you done well in the face of adversity? Have you spent time in another culture? A leader? An athlete? How have these experiences made you better, and how will your skills help you to succeed and stand out amongst a group of other high-achieving peers? If you think it is relevant, tell me about it. Don't have anything like this to write about? That's OK too – maybe instead you can write about what challenges you expect to face in the transition to university and how you will overcome them.'
>
> Barry Sullivan, Head of Admissions,
> Cardiff School of Engineering, Cardiff University

Joint honours courses

Personal statements for joint honours courses are usually read by selectors from both of the departments to which you are applying. If you want to apply for a joint honours engineering and management course, someone from the engineering faculty will expect to read about engineering while their colleagues from the management department will want to read about management. This is fine if you apply to five similar courses, but if you apply for, say, three engineering with management and two single honours engineering courses, you will find it difficult to satisfy the selectors from the two different types of courses. Discuss your course choices with your careers adviser to assess how best to present these choices in your personal statement.

Applying for different courses at the same university

You can apply for more than one course at a particular university; you do not have to choose five different universities. But applying for two courses at the same university does not necessarily increase your chances of studying there. Take the case of a student who is desperate to study at a particular university, perhaps because she has friends there or she likes the city. She decides to apply for both the civil engineering and biomedical engineering courses. The admissions tutor for civil engineering will look at her application for the civil engineering course, and the admissions tutor for biomedical engineering will look at the application for this course.

Our applicant's main interest is civil engineering, so her personal statement emphasises this, but it also devotes one paragraph to her interest in biomedical issues. The civil engineering admissions tutor reading the personal statement will judge it on how it addresses this course, so he or she might not be fully convinced that the student is serious because the personal statement will not focus enough on the reasons for the choice of civil engineering and what the candidate has done to investigate it (reading, work experience etc.). Similarly, biomedical engineering is a very specialised area, and the admissions tutor for this course will expect to read a personal statement that focuses on this, and he or she is not going to be very interested in reading about bridges and roads.

So, by trying to give herself a better chance of getting to this university, the applicant is actually reducing her chances. There are some instances where it is possible to apply for two separate courses at the same university if they are very similar, but it would be advisable to discuss this with the university admissions department before doing so. So, when writing the personal statement, try to imagine how it will come across to each of the departments to which you are applying. Do not try to write something too general in order to allow yourself the luxury of applying to a wider range of courses.

The structure of the personal statement

For students applying to start their studies in 2026, personal statements are changing from one longer piece of text to three separate sections, each with a different question to help shape the focus for students' answers. Each section will have a minimum character count of 350 characters, which is clearly labelled on the question boxes, along with an overall character counter to ensure students know if they're on track. The new web page for submitting the personal statement will also feature helpful on-page guidance for each question.

The new questions are as follows:

1. Why do you want to study this course or subject?
2. How have your qualifications and studies helped you to prepare for this course or subject?
3. What else have you done to prepare outside of education, and why are these experiences useful?

Responding to each question

1. Why do you want to study this course or subject?

This could include:

- what first got you thinking about engineering, for example, watching the news about a new engineering project, an article in a newspaper about new developments in the mobile phone industry or personal experience such as work experience? This could date back many years, for example, how you liked taking things apart and reassembling them when you were very young;
- things you have studied at school, for example, a topic in physics or chemistry that particularly interested you;
- how your particular interests and qualities make engineering a suitable career.

2. How have your qualifications and studies helped you to prepare for this course or subject?

This could include:

- academic achievements; for example, Maths Challenge or the Mathematical Olympiad;
- extracurricular activities and accomplishments;
- responsibilities; for example, sports captain, head of house, chairman of a club or society;
- hobbies.

3. What else have you done to prepare outside of education, and why are these experiences useful?

This could include:

- books, articles, magazines or websites that you have read (anything here must be contextualised and emphasise what you have learnt);
- work experience (see Chapter 2);
- talks or lectures;
- school visits to museums or exhibitions;
- the relevance to engineering of things you have studied at school;

- Extracurricular activities; for example, sports, school orchestra, Management Challenge, Duke of Edinburgh expeditions;
- The Extended Project Qualification.

Other information relevant to the application

This could include:

- gap year plans;
- if you are an international student, why you want to study in the UK;
- what qualities or experiences you could offer to the university or the course.

Blogs and online lectures

Since the Covid-19 global pandemic, virtual open days and tours have become commonplace. Technology has come to the rescue with strong, interactive platforms available to roughly replicate the normal experience of university teaching. Many universities upload lectures so that they are available to the public (including potential students). If you are unable to go to talks and presentations because of school commitments or because you do not live near a university that offers these, then downloading online lectures or following blogs is a good way to learn more about engineering.

Two good examples (Imperial College London and the University of Bath) are given below www.imperial.ac.uk/be-inspired/social-and-multimedia/ and https://soundcloud.com/uniofbath/sets/public-lecture.

Personal statement writing with Artificial Intelligence (AI)

While using GenAI, such as ChatGPT, to write your personal statement for you could be considered cheating, there are effective ways to use AI as a tool to help you brainstorm, structure and refine your personal statement. The key is remembering that your personal statement should reflect *you* – your experiences, skills and aspirations. AI can assist, but it should never replace your own thoughts.

Acceptable use of GenAI in developing your personal statement:

Brainstorming ideas

GenAI can be a great starting point for brainstorming topics relevant to the course you're applying for. For example, you can ask ChatGPT about skills or qualities that are valued in the subject you're interested in, then think about how these relate to your personal experiences. This helps you generate ideas without directly copying the suggestions, ensuring your statement remains authentic.

Structuring your personal statement

If you're unsure about how to structure your personal statement, GenAI can suggest a clear outline or format. It can help break down the content into logical sections – introduction, academic achievements, relevant experiences and future goals – making it easier for you to organise your thoughts.

Improving readability

Once you have a draft, GenAI can be useful for improving the readability of your personal statement. It can suggest ways to simplify sentences, remove redundancy or make your writing more concise without altering the meaning, making sure your statement is clear and engaging.

Universities value the authenticity and personal touch in your application. Use AI wisely to guide your process, but always make sure the final draft is entirely your own, reflecting your unique experiences and aspirations.

Preparation for writing the personal statement

1. **Evidence of study of what the subject involves.** This can be anything from indicating some relevant books or websites to having gained work experience with a local company. Good websites, such as the Royal Academy of Engineering, have a lot of useful and reliable explanations about what engineering involves. Increasingly, applicants have been attracted to engineering by watching various programmes on the Discovery Channel, such as *Extreme Engineering*; this is fine.
2. **Courses and summer schools.** A student who has made the effort to attend a summer school or Headstart course has demonstrated that they have tried to find out what a degree involves. This isn't necessarily the same as what a career involves, but is a good start.
3. **Evidence of an understanding of the breadth of the subject.** Most engineering disciplines are associated in the popular imagination with specific elements, for example, mechanical engineering with Formula 1, or civil engineering with structural design. Any evidence of understanding that mechanical engineers also study thermodynamics, efficient engines, sustainable design and so on, or that civil engineers are particularly concerned with water treatment, flow and management, soil mechanics and geotechnics and are involved in people management, will demonstrate some depth to the background research.

4. **Passion.** It might be argued that an applicant doesn't know enough about the subject to yet be passionate about it. This might have been true 20 years ago, but the internet and YouTube have put an end to that. Any applicant interested in engineering can find a wealth of informative resources that could excite their interest.
5. **Understanding the value of transferable skills.** Sometimes applicants approach the exercise in too narrow a fashion. Usually this manifests itself as a focus on the technical side of engineering and demonstrates a lack of awareness of the broad transferable skills that are needed and developed in a degree. Accordingly, almost any work experience can offer valuable insights into the applicant's approach to team-working and problem-solving. It will not be enough to mention in some vague way that you like working in a team; it would be far better to explain how a specific team-based activity has helped you to understand your strengths and weaknesses.
6. **The value of work experience.** Often applicants are reluctant to mention work experience unless they think it is directly relevant. I have known applicants not mention a Saturday job because it is in a supermarket. However, any job demonstrates experience of the world of work and offers opportunities for the curious student to find out how the business works, even if they are just shelf-stacking.
7. **Studying A levels or equivalent requires commitment.** So, any applicant who is managing to do well at this level and is still maintaining committed contact with organisations such as Scouts, local sports teams or musical productions, for example, is demonstrating the necessary ability to work hard and play hard that is required at university.
8. **Think very carefully about what to write about leisure activities.** You have to remain honest, but if all you do is play games and socialise, now might be a good time to start broadening your activities. Some applicants make the effort to go to local branch meetings of professional engineering societies, others subscribe to industry magazines. All such things demonstrate some commitment to the career. Remember, you are only at university for three or four years; your engineering career is for life. You need to expose yourself to the industry in some way to find out what the job involves.

Lawrence Coates, Professor of Engineering,
University of East Anglia

Advice from a student

'When writing a personal statement for your UCAS application, my advice would be to first write everything you can think of that is remotely relevant, including every reason you have for wanting to study your chosen subject and all the extracurricular activities you've done. From there, you can begin to redraft and remove the things you think are less useful to your application. Bear in mind that the majority of your personal statement should be focused on academic issues and only a paragraph should be given to personal hobbies, such as sport and music. When talking about non-academic activities, you should try to explain the skills that this activity has helped you develop. For example, if you play a musical instrument you can explain that it has helped you gain personal discipline – you should explain to the reader how your activities have made you into a better university candidate. Ensure your personal statement is true and that you've read everything you said you had. You should have notes on the books you've read which you can use to refresh yourself on their contents before your interview so that when you are asked about it you have something intelligent to say.

'When preparing for an interview you should always do a mock interview with someone you don't know well to best simulate the environment in which you will find yourself. Also, to prepare for an interview, you should write out model answers to some of the stock interview questions, such as "Why do you want to study this subject?", "Why do you want to study at this university?" and "Why are you suited to study this subject?" Remember, you should not rush your UCAS application, so start preparing as early as you can to maximise your chances.'

Jordan Massiah, Magdalene College,
University of Cambridge, MEng Engineering

Analysing a personal statement

Sample personal statement

1. Why do you want to study this course or subject?

I first became interested in engineering during my A level Physics classes. As a child, I had always enjoyed taking things apart, although I was not always able to put them back together again. This early fascination developed into a deeper interest in engineering, and I have since been keen to explore how engineering works in practice, particularly within the fields of architecture and construction.

2. How have your qualifications and studies helped you to prepare for this course or subject?

Alongside physics, I am studying Mathematics and History of Art. Mathematics is an important tool for engineers, and it has taught me to think in a logical, problem-solving way, which I know will be valuable in engineering. History of art, while seemingly unrelated, has actually proved to be very useful in developing my analytical skills. It puts architecture into a social and political context, as well as exploring how materials have influenced the development of both architecture and sculpture. This combination of subjects has given me a broad skill set, blending technical problem-solving with creative and analytical thinking. I believe that my combination of A levels and my research into engineering as a career make engineering an ideal choice for me. The experiences I've had have all contributed to a solid foundation that has prepared me to further my studies in this field.

3. What else have you done to prepare outside of education, and why are these experiences useful?

To further investigate engineering, I spent a week at a local architecture practice, where I saw how structural engineers collaborate to ensure their designs are practical. I also visited a construction project as part of my school's work experience programme, and it was fascinating to see how theoretical concepts I had studied were applied in real-life situations. In addition to these experiences, I've read books like *A Short History of Engineering Materials* by John Cameron, which has deepened my understanding of the materials used in engineering. I've also attended public lectures on engineering and had conversations with engineers, which has given me a much clearer idea of what a career in engineering entails.

Points raised by this personal statement

An admissions tutor who read this sample personal statement made the following points.

General

While it is clear that the candidate has done some research, there is very little detail in the statement – it is very general – and so I do not really get a clear picture of the depth of knowledge the candidate has about engineering or about his/her particular areas of interest. It is also, to be honest, a little on the bland side and also a bit frustrating – a lot of sentences that should lead to something that will interest me end without giving me any information.

Specific points

1. Why do you want to study this course or subject?

It would have been nice if they could have given an example – perhaps it was electricity, the behaviour of materials or some problems involving forces? This is the students opportunity to express their passion for their chosen career path. Further depth is required here.

2. How have your qualifications and studies helped you to prepare for this course or subject?

More detail on the relevance of the students academic experience would have strengthened this section. The candidate could have told me about the skills they have learnt in their subjects and their relevance to an engineering degree.

This would have been an ideal opportunity for the candidate to show me that they have been really thinking hard about their future career and about whether they have the right qualities to be a successful engineer.

Not many applicants for engineering study history of art, so this immediately makes them stand out. And I don't know much about the history of art, so an example here would be interesting for me to learn about – I'm sure it would make me want to learn more by meeting the candidate.

3. What else have you done to prepare outside of education, and why are these experiences useful?

This should have been the most interesting part of the statement. The candidate should have gone into more detail about a specific experience that captured their imagination. With regards to the students wider reading, I am always encouraged when students read around the subject, but what I would like to know is more detail. How do the ideas in the book link to A level study and the real world?

A revised (and much better) version of the personal statement based on the above advice is now given.

Revised sample personal statement

1. Why do you want to study this course or subject?

I first became seriously interested in engineering during my A level Physics classes when we looked at the properties of solid materials, and I began to understand why, for example, the development of reinforced concrete revolutionised the construction industry. As a child, I had always enjoyed taking things apart, although I was not always able to put them back together again. This early fascination developed into a deeper interest in engineering, and I have since been keen to explore how engineering works in practice, particularly within the fields of architecture and construction.

2. How have your qualifications and studies helped you to prepare for this course or subject?

Alongside physics, I am studying mathematics and the history of art. Mathematics is an important tool for engineers because the starting point of any engineering project is an analysis of its feasibility. The use of integration to find the centre of mass, for example, can help with the design of an asymmetric building. The history of art is an analytical subject and puts architecture within a social and political context. It also involves looking at the use of material and how architecture and sculpture were able to develop as new materials were introduced. The transition from the small, squat Romanesque churches to tall and graceful Gothic cathedrals in Europe was due to the invention of flying buttresses, which, in turn, were only effective when tensile forces were reduced by the addition of heavy statues or decorative stone elements. This combination of subjects has given me a broad skill set, blending technical problem-solving with creative and analytical thinking. I believe that my combination of A levels and my research into engineering as a career make engineering an ideal choice for me. The experiences I've had have all contributed to a solid foundation that has prepared me to further my studies in this field.

3. What else have you done to prepare outside of education, and why are these experiences useful?

To further investigate engineering, I spent a week at a local architecture practice, looking at how they worked with a structural engineer to ensure that their designs were practical. I also visited a construction project as part of my school's work experience programme, and it was fascinating to see how theoretical concepts I had studied were applied in real-life situations. The engineers explained that the concrete girders used to hang the curtain walls of the office block had to be strengthened along their top surfaces because in a cantilever the tensile forces are at the top, and concrete is weaker in tension than it is in compression. I've also read as much as I can about engineering, including *A Short History of Engineering Materials* by John Cameron, and I was fascinated by how the use of cast iron in early 20th-century American enabled architects and engineers to build the prototypes of today's skyscrapers. I've enjoyed attending public lectures on engineering and talking to engineers, who have given me a much clearer idea of what a career in engineering entails. I began to understand that an engineer needs to be able to analyse information quickly and to be able to solve problems. My aim is to study structural engineering and then work alongside architects in the creation of exciting new buildings.

I enjoy sport and music. I am captain of my school football team and so have had to develop leadership and communication skills, as well as physical fitness. I play the guitar in a band and the cello in the school orchestra, both of which help me with my manual dexterity and teamwork. Outside of school, I enjoy cooking and cycling. I am a member of my local cycling club and compete most weekends.

Adding the extra information requested by this admissions tutor would add detail, make it more interesting for him to read (so he is more likely to want to meet the student), demonstrate that the student is interested enough in the subject to be thinking about links between his studies and what he has experienced, and bring it up to the required length.

Linking your interests and experiences

In your personal statement, try to avoid creating what amounts to a list of things you have read, studied or experienced. It is better to make connections between these in order to demonstrate that you have thought carefully about what is required to be a successful engineer:

- how your A level studies are related to things you will study at degree level;
- how something you observed during work experience stimulated further reading;
- how skills that you gained from extracurricular activities, such as communication or leadership, are useful for potential engineers.

You could link:

- an article you read about increasing the battery life of mobile phones with something you studied about chemical reactions in A level Chemistry;
- the design of a new building with your study of forces in physics;
- the need for engineers to be good communicators with your role as your class representative at school;
- the follow-up research you did on wind turbines with an on-site visit on a school trip;
- a news story on a new aircraft design with an article in the *New Scientist* about composite materials.

Work experience

Work experience is important as it demonstrates a commitment to the subject outside the classroom. Remember to include any experience, paid or voluntary. If you have had relevant work experience, mention it on your form. Explain concisely what your job entailed and what you got out of the whole experience. Even if you have not been able to get work experience, if you have spoken to anyone in engineering about their job, it is worth mentioning, as all this information builds up a picture of someone who is keen and has done some research. See Chapter 2 for further information.

How to get started on the personal statement

Different people approach this in various ways. If you are struggling to get started, begin at the end or in the middle. To start at the end is often a good way to write this statement, as the summation should be your sales pitch as to why you want to take this course. If you know that, you can work your way through the rest of the statement. If you treat these as isolated paragraphs, then you will be able to access them more easily. Another good strategy is to start by making lists of anything that you think is relevant to your application. Then begin to organise them into sections. Your personal statement could include some of the following points.

I first became interested in engineering because . . .

- I read an article in a newspaper about . . .
- I read the book '. . .'
- I saw a piece on the news about . . .
- of my work experience;
- of my father's job;
- of something I have enjoyed studying at school. I have investigated engineering by . . .
- reading books;
- reading the *New Scientist*;
- reading the Royal Academy of Engineering website;
- work experience;
- going to a public lecture at a university;
- discussing engineering with an engineer;
- downloading a podcast of a university lecture.

From my work experience, I learnt . . .

- that the qualities necessary to become a successful engineer are . . .
- how the theory we study at A level is applied practically;
- the importance of communication skills/problem-solving/leadership.

Other points to include:

- a particular A level topic is useful because . . .
- my part-time job is useful because . . .
- my role as rugby captain has taught me . . .
- being leader of the school orchestra has taught me . . . (or lead in the school play, or . . .)
- during my gap year, I will be . . .
- I was awarded first prize for . . .

Only when you have the ideas listed should you start to write full sentences, including them in the relevant sections of the personal statement structure.

Many people find that the best way to make it fit the available space is to start with a statement that is perhaps double the allowed word count but that incorporates everything you think is relevant, and then start to edit it down by:

1. removing any superfluous adjectives or complicated sentence structures, for example, 'I was lucky enough to attend a stimulating talk on the exciting topic of wind turbines delivered by the renowned engineer Richard Martin, which inspired me to read xxx', could more succinctly be written as, 'I read xxx after hearing a talk on wind turbines by Richard Martin';
2. removing any passages that are not about you. Often, I read personal statements that try to impress the admissions tutors by explaining in detail engineering concepts that they have read about. The chances are that the person reading the statement is far more familiar with the topic than you are, so will skip this section completely, and so you have wasted valuable space that could have been used to tell him/her more about you;
3. removing sentences that state the obvious: 'I chose physics and mathematics at A level because these will be useful when I study engineering';
4. removing passages that contain nothing factual, particularly closing statements such as: 'I believe that my commitment to engineering coupled with my academic achievements and ability to work hard will make me an asset to your esteemed institution'.

Once you have gone through this process, you should be getting near a personal statement that is around the right length. Then you can fine-tune it so that it fits exactly. A useful tip is to ask your referee to include anything that you have had to discard in his/her reference. So, if you have run out of space and have to omit information about, say, your role in organising a school event, then the referee can incorporate this into the Reference section of the form.

Things to avoid at all costs

- Boasting about your achievements.
- Telling the admissions tutor why he/she should offer you a place.
- Quotes from famous engineers, scientists or philosophers.
- Listing your examination results or predicted grades (these appear elsewhere on the UCAS application).
- Adapting or copying personal statements that you have found on the internet or in books (including this one!).
- Spelling mistakes or grammatical errors.

First impressions

Engineers tend to be practical people who can convey complicated information in ways that are easy to understand. They are problem-solvers who can get to the heart of an issue, discarding information that is not relevant to the issue they are dealing with. They understand the importance of turning ideas into solutions that have a benefit to others. Bear this in mind when writing the personal statement. Make the statement easy to read by using simple sentences, adding spaces between paragraphs and sticking to the point. Give it a coherent structure and a logical flow. Remove anything that does not have a practical use. Treat the statement as you would an engineering project. Its purpose is to get you the offer at your chosen university.

Language

It is important that you use succinct language in your personal statement and make every word count. Remember that you are limited on the number of characters you may use, so it is important not to use up this vital space with superfluous language. Keep it as simple and clear as you can, rather than using overcomplicated language in an effort to impress.

For example:

I was privileged to be able to undertake an internship with a well-known engineering company where I was able to see the benefit of having the ability to be confident with information technology – approximately 200 characters.

Could be rewritten as:

My three weeks' work placement at Rolls-Royce showed me the importance of being proficient in using spreadsheets – approximately 110 characters.

Similarly:

I was honoured to be chosen to play the lead role in my most recent school drama production, and interacting with the producer and the rest of the cast involved a significant degree of communication and teamwork – approximately 210 characters.

Could be rewritten as:

Performing as the lead in my school play taught me to work and communicate effectively with others – approximately 100 characters.

Phrases to avoid

- 'It was an honour to . . .'
- 'I was privileged to . . .'
- 'From an early age . . .'
- 'For as long as I can remember, I have dreamt of . . .'

Writing the personal statement

- **Accuracy of spelling and grammar.** The current jobs market is such that companies are using software that automatically rejects applicants after three spelling mistakes. So it is important to demonstrate this attention to detail in a personal statement. The statement should not have any mistakes at all when it is finally polished and submitted. This probably means it will have been through 20 draft versions. It conveys a lot to an admissions tutor when an applicant has made a spelling or grammatical error, or has doubled or missed out words. If engineers don't pay attention to detail, there could be serious repercussions in most engineering disciplines, including safety issues.
- **Written communication.** Most practising engineers spend a lot of their daily working life writing good English in technical reports. Most applicants study mathematics, physics and chemistry and hence do not have formal opportunities to improve their written English. So, any evidence that an applicant has attempted to broaden their reading beyond textbooks or involved themselves in extended writing will be useful.
- **Structure.** Some personal statements are chaotic and rambling. Good ones have obviously been planned and tell a story about the applicant in a logical order. It is important for an applicant to realise how much time it takes to prepare a statement.
- **Peer review.** An essential step in constructing a personal statement is to let a close friend look at it. Frequently they will notice some important aspect of your life that you have forgotten to mention because it is so obvious.
- **The personal statement can steer the interview.** Some universities use the personal statement as a basis for allocating applicants to staff for interviews. So it is important to only include things that you want to discuss in these circumstances. So, including things because you think it will look good, but not knowing much about them, is a bad idea.

> - **The personal statement must be personal.** It is usually quite obvious if an applicant is trying to guess what the reader wants. It is far better to write from the heart, demonstrating passion for the subject. So many personal statements look the same. So, when an admissions tutor reads one with some interesting element that nobody else has mentioned it livens it up. Start by thinking of five things that could only be included in *your* personal statement; in other words, make sure it is personal not formulaic.
>
> Lawrence Coates, Professor of Engineering,
> University of East Anglia

Sample personal statements

The following examples of personal statements have been contributed by students who were successful in gaining offers from their chosen universities. They have been adapted to follow the new three-question format.

Please remember that these are personal statements; that is, they reflect the experiences and ambitions of the students who wrote them – so do not attempt to copy them or adapt them for your own use.

Personal statement 1

1. Why do you want to study this course or subject?

My connection to engineering stems from my roots in Khaibakh, a village in Chechnya that was razed in 1944, notorious for the massacre of over 700 Chechens deported by the Stalinist regime. The village was rebuilt over time, and this process sparked my interest in how engineering can restore communities and preserve history. Growing up, I observed the engineering involved in these rebuilding efforts, particularly the techniques used in the construction of the village tower by my ancestors. It was fascinating to consider how modern materials and methods contrasted with the older structures while respecting their history.

This experience fuelled my desire to study civil engineering. I'm drawn to the idea of combining scientific curiosity with the ability to create structures that positively impact society. Engineering is not just about building; it's about understanding the people, culture and history behind the structures we create. I believe civil engineering will allow me to make a difference through thoughtful, innovative designs that respect both tradition and progress.

Civil engineering perfectly blends my scientific curiosity and my desire to create solutions that positively impact society. Through my education and practical experiences, I have gained a deeper understanding of the complexities involved in engineering. I am excited to continue developing my skills and contributing to the creation of structures that respect both the past and the future, serving the communities that rely on them.

2. How have your qualifications and studies helped you to prepare for this course or subject?

My educational journey has blended scientific curiosity with an appreciation for people and culture. Moving to the UK allowed me to improve my language skills, understanding that effective communication and cultural awareness are essential for designing structures that serve communities.

In my A levels, I focused on physics and mathematics, laying the foundation for my engineering studies. I further developed my knowledge by reading *To Engineer is Human* by Henry Petroski, which emphasised how failure informs engineering success. This concept became clear during my summer placement at BMCE, where I worked on a car park expansion project. I sketched cross-sections of the structure to help clients visualise it, applying my physics and geometry knowledge to understand materials and structural design. I also read *A Short History of Engineering Materials* by John Cameron, deepening my interest in how cast iron was used in early skyscrapers and inspiring me to learn more about the materials that build our world.

3. What else have you done to prepare outside of education, and why are these experiences useful?

Beyond my studies, I sought practical experiences in engineering. At BMCE, I worked on the addition of two steel floors to an existing car park, gaining experience in reading engineering plans, analysing materials and understanding structural integrity. I also visited construction sites where I saw the application of theoretical concepts, such as calculus, to real-world issues. One site visit involved a residential building with cracks in its foundations caused by expanding tree roots, highlighting the need for engineers to consider environmental factors in design.

Additionally, my project management role at Cummings Group exposed me to the business side of engineering. I was involved in a £70m student accommodation project that escalated to £160m when a Shakespearean theatre was discovered during excavation. This taught me how unexpected discoveries can impact budgets and timelines. I attended site meetings, learning about safety regulations, scheduling and how additional costs affect projects. This experience reinforced the importance of understanding both technical and practical aspects of engineering.

Personal statement 2

1. Why do you want to study this course or subject?

My passion for engineering began at a young age, starting with Meccano and progressing to more complex projects. By age 10, I was racing and maintaining my own Rotax go-kart. At 11, I had already performed major upgrades on a 22-year-old jet ski, modifying the supercharger and adding an intercooler to increase power. Since then, I've worked on numerous vehicles, including a Jeep JK Wrangler, a Dodge Ram 1500 pickup and a Mercruiser 5.7 L V8 speedboat engine. Most recently, I built a modified 1986 BMW 3 Series, installing a 2.5 L twin-turbo Toyota 1JZ-GTE engine and adapting its internals. These projects ignited my desire to pursue a degree in Automotive Vehicle Design and Performance. I am particularly eager to gain hands-on experience with CAD tools such as CATIA and Ansys, which I have read about but am excited to use in practice.

My fascination with engineering extends beyond vehicles. I am also captivated by the idea of space tourism, sparked by Sir Richard Branson's ventures. The possibilities in space travel highlight how engineering pushes the boundaries of what we think is possible. The history of the land speed record, for example, demonstrates how engineering constantly evolves. In 1904, William Vanderbilt broke the record at 92 mph, but today, Bloodhound SSC aims to reach 1,000 mph with a car powered by a Eurofighter Typhoon engine. This progression demonstrates not only engineering's innovation but also humanity's adaptability and drive to achieve the seemingly impossible. I want to be part of this dynamic, ever-evolving field and contribute to the engineering feats of tomorrow.

Automotive Vehicle Design and Performance offers the perfect blend of my practical experience, academic interests and passion for innovation. The combination of engineering theory and real-world application excites me, and I am eager to further develop my skills with the advanced tools and technologies used in the industry. With a strong foundation in mathematics, physics and hands-on engineering experience, I am well-prepared to embark on this course.

2. How have your qualifications and studies helped you to prepare for this course or subject?

After completing my A levels, I was accepted onto the Engineering Foundation Programme at City University. This programme has allowed me to build on my A level knowledge and apply it to real-world engineering problems. Through studying heat transfer, I became interested in thermofluids and entropy, topics I look forward to exploring in more detail at the undergraduate level. My studies in physics have deepened my understanding of materials used in design, while mathematics has prepared me for the programming elements essential in modern engineering. These subjects have strengthened my analytical and problem-solving abilities, preparing me for the academic challenges of an automotive engineering degree.

5| The Personal Statement

3. What else have you done to prepare outside of education, and why are these experiences useful?

To further develop my understanding of automotive engineering, I have undertaken internships that allowed me to apply my self-taught skills in professional settings. In 2022, I worked at a Mercedes Service Centre and Repair Shop in Cairo, where I gained hands-on experience in repairing and servicing cars. I learnt important skills such as topping up AC systems with Freon, adjusting ignition timing and servicing both manual and automatic transmissions.

In 2023, I had the opportunity to work at Aston Martin's state-of-the-art service centre in Dubai. There, I memorised the specifications of various models, learnt wheel alignment and diagnostics and balanced wheels with carbon ceramic brakes. Most importantly, I gained a deeper understanding of the teamwork required in automotive engineering. I learnt that every aspect of a car's design and service must come together seamlessly, and I was commended for my contribution to the team. These experiences taught me the importance of collaboration and technical expertise, qualities essential to success in engineering.

Personal statement 3

1. Why do you want to study this course or subject?

From my early days with Mindstorms robots in middle school to building a hydroelectric plant model, engineering has always been a way for me to overcome challenges. I started with simple robot designs, like rover robots and those solving Rubik's cubes. These projects sparked my interest in engineering and problem-solving. Over time, I developed a curiosity for mathematics and physics, particularly the Golden Ratio and its real-world applications in nature and architecture, which became the topic of my IB research paper. Reading *Dimensional Analysis* by Don S Lemons deepened my understanding of how dimensionless properties, such as Reynolds and Nusselt numbers, explain physical phenomena. This theoretical knowledge complements my hands-on experiences, such as modifying motors and designing mechanical systems. I want to study engineering to apply these principles in creating real-world solutions, especially in energy efficiency and sustainable technology.

Engineering excites me because it combines my love for science and mathematics with the opportunity to create practical innovations. My involvement in Olympiads, independent research, engineering competitions and internships has equipped me with skills in problem-solving, leadership and teamwork. These experiences have prepared me to tackle the challenges of modern engineering, and I am eager to contribute to innovative solutions that address global issues.

2. How have your qualifications and studies helped you to prepare for this course or subject?

My academic studies in mathematics and physics have provided a solid foundation for engineering. I have excelled in these subjects and sought opportunities to explore engineering further. Participating in Olympiads like the UK SMC, Senior Physics Challenge and UKBC taught me to approach complex problems methodically and apply my knowledge in new contexts. For example, after watching *The Man Who Knew Infinity*, I was inspired to research series and discover a recurrence formula for discrete sums of powers, deepening my understanding of mathematical theory.

At the Sirius Education Centre, I worked on building a hydroelectric plant model, bridging theoretical knowledge with practical application. There, I also gained experience in CAD modelling and 3D printing, which I applied in engineering competitions like the First Tech Challenge (FTC) to develop robots. These experiences enhanced my problem-solving abilities and prepared me for the multidisciplinary nature of engineering.

3. What else have you done to prepare outside of education, and why are these experiences useful?

Beyond academics, I've sought real-world engineering experience through internships at Tecom Group Inc. and Euroclima. At Tecom Group, I learnt how broadcasting signals are monitored with detectors, and at Euroclima, I observed how air handling units (AHUs) are designed to meet customer specifications while maintaining performance. These internships exposed me to fluid mechanics, advanced circuits and dimensionless numbers in practical applications, deepening my understanding of the real-world relevance of engineering.

As the manufacturing team leader in the First Robotics Competition (FRC), I applied CAD and 3D printing skills to solve technical challenges. My team and I collaborated to design and construct a robot that met specific performance criteria while troubleshooting unexpected malfunctions. This experience taught me valuable lessons in teamwork, creativity and adapting designs to ensure project success. The FRC competition also reinforced my desire to pursue engineering, where I can continue learning, adapting and innovating.

5| The Personal Statement

General tips for completing your UCAS application

- Before submitting it, also ensure you check your application very carefully for careless errors that are harder to see on screen.
- Keep a copy of your UCAS application so you can remind yourself what you wrote prior to an interview.
- Ensure that you have actually done all the things you mentioned in the statement by the time you are interviewed.
- Research the full course content, not just the first year.
- Research the entry requirements.
- Ask your teachers for your grade predictions.
- Ensure your personal statement is directed at the courses you are applying for.
- Include lots of detail in the personal statement.
- Get someone else to proofread your personal statement.
- Illustrate your points with examples and evidence.
- Do not waste valuable space in the personal statement – make every word count.

Joke

Two antennas got married – the wedding was lousy, but the reception was outstanding.

6 | A well-oiled machine: Succeeding at interview

Not all universities interview candidates, but it is likely that at least one of your five choices will do so, so you need to be prepared. Your interview will decide whether you will be offered a place or not. The information on the UCAS application will have been the basis on which the decision to interview you was made, but a good UCAS application cannot help after a poor interview. So prepare thoroughly. Here are some important points to consider.

- If you interview well, and you subsequently narrowly miss the grades that you need to take the place, you may still be offered the place.
- Interviews are normally conducted in an informal and relaxed manner, the purpose of which is to allow you to talk about your interests and suitability for the course.
- Think about the impression you will make – think about your body language, eye contact and communication skills.
- Go into the interview with a mental checklist of what points you wish to mention and try to steer the interview to address these (see below).
- Interviewers are less interested in investigating your subject knowledge than in looking at how suitable and committed you are for their course. So, evidence of research and appropriate qualities such as analytical or problem-solving skills are important elements of a successful interview.
- Remember, your future teacher or lecturer might be among the people interviewing you. Enthusiasm and a genuine commitment to your subject are extremely important attitudes to convey.
- An ability to think on your feet is vital – engineering is about problem-solving. Don't try to memorise potential answers or responses, but be prepared to expand on things you have mentioned in the personal statement.
- Important preparation includes rereading your UCAS personal statement. Never include anything in your UCAS application that

you are not prepared to speak about at greater length or in more detail at the interview.
- Questions may well be asked on your extracurricular activities. The interviewer may do so either to put you at your ease or to find out about the sort of personal qualities you possess; therefore, your answers should be thorough and enthusiastic.
- At the end of the interview, you may be asked if you have any questions. Often, this is simply a polite way of ending the interview, so do not feel that you need to ask anything. Just say, 'Thank you, but all my questions were answered during the introductory lecture today and by the students who showed us around. If I think of anything I will contact the admissions department.'
- However, it can also sometimes work in your favour if you do ask a question, as long as it is not something that you could find out for yourself. Try and tailor any questions around the subject itself, showing your genuine enthusiasm for the subject or maybe pick up on something you were not able to answer during your interview. This can be another way of demonstrating your willingness to learn.
- Try and explain a topic to a friend or family member who knows nothing about the subject. This is an exercise in communication skills and your ability to clearly exemplify your point.
- Finally, smile, thank them and shake hands. Above all, convey your enthusiasm so that they will remember you at the end of a long day of interviews.

'Within the Department of EEE, we actively encourage all applicants that are expected to meet our very high entry standards to attend their interview afternoon. The reason is not to test their academic ability, but to ensure that they have what it takes to succeed at Imperial in their studies. Perhaps more importantly, we want our high-calibre students to be happy during their stay at Imperial and so it is important to know that they want to study EEE for the right reasons. To this end, during their interview, we are looking for a student who is "switched on" (as opposed to an applicant having been primed for the occasion) and genuinely keen to pursue a career in EEE (it is not uncommon for parents to push their children into this subject). We ask each applicant a few technical questions, of increasing difficulty, to gauge their limits of mathematics and physics; not necessarily related to EEE. In addition, we are looking for a confident, inquisitive mind, as well as signs of weakness areas. Since places are limited, we try to find signs that the applicant will have a long-term commitment to academic life, as well as indicators that suggest that they will help to enrich the lives of those around them.'

Dr Stepan Lucyszyn, Undergraduate Admissions Tutor,
Department of Electrical and Electronic Engineering (EEE),
Imperial College London

Preparing for an interview

If you are called to interview, it is advisable that you reread your personal statement to remind yourself of what you wrote, so that you can be ready to answer questions on it at interview. Preparation for an interview should also be an intensification of the work you are already doing outside class for your A level courses. Interviewers will be looking for evidence of an academic interest and commitment that extends beyond the classroom. They will also be looking for an ability to apply the theories and methods that you have been learning in your A level courses to the real world.

Essentially, the interview is a chance for you to demonstrate knowledge of, commitment to and enthusiasm for engineering. The only way to do this is by trying to be as well informed as you can be. Interviewers will want to know your reasons for wishing to study engineering, and the best way to demonstrate this is with examples of things you have seen, read about or researched. Later in the chapter, there is a section on current issues that you can use to kick-start your reading.

Newspapers and magazines

Before your interview, it is vital that you are aware of current affairs that relate to the course for which you are being interviewed. *New Scientist* will give you a good grasp of scientific and engineering developments, as will reading the science sections of newspapers. You should also keep up to date with current affairs in general.

Magazines can be an important source of comment on current issues and deeper analysis. There are many specialist engineering publications, such as *The Engineer, New Engineer, Structure* magazine and *Aviation Week*. Further details can be found in Chapter 11.

Television and radio

It is also important to watch or listen to the news every day, paying particular attention to reports on scientific and engineering issues. Documentaries, programmes and podcasts about engineering projects can be enormously helpful in showing how what you are studying is applied to real-world situations. Look out for television programmes that showcase innovations in engineering, ranging from those aimed at a general audience, such as *The Great British Railway Journeys* or *Grand Designs*, to more specialist and factual series like *Horizon, Engineering Giants, Building the Impossible, Mega Manufacturing* and *Massive Engineering Mistakes* on platforms like BBC, Netflix and National Geographic. More recent titles, including *The World's Most Extraordinary Homes* and *Impossible Engineering*, highlight the creative and technical challenges behind iconic structures. Radio series such as BBC Radio 4's *Inside Science, The Life Scientific, The Curious Cases* of *Rutherford & Fry* and *52 Minutes of Science* are also valuable for understanding engineering breakthroughs and their societal impacts.

The internet

A wealth of easily accessible, continually updated and useful information is, it goes without saying, available on the internet. Given the ease with which information can be accessed, there is really no excuse for not being able to keep up to date with relevant current issues. Online platforms give free access to thousands of lectures and presentations from universities around the world; newspapers can be read online . . . the list is endless. In this age of information overload, anyone who is serious about keeping abreast of current issues (or wants to be seen as being serious) has unlimited opportunities to do so. Thus, an interviewer is not going to be impressed with a student who claims that he or she has been too busy to know what is happening in his or her chosen areas of interest.

- Check online news websites every day to read the latest news stories.
- If you cannot buy a newspaper every day, look at an online version, for example, www.theguardian.com.

Podcasts

Try to listen to engineering- and science-related podcasts to keep up to speed with the latest news, innovations and ideas within the field. Series such as *The Engineering Leadership Podcast*, *The Structural Engineering Channel* and *Overcoming Challenges in Engineering* offer practical insights for aspiring and practicing engineers. Broader science and innovation podcasts like *The Curious Engineer, Stuff You Should Know, 50 Things That Made the Modern Economy* and *How I Built This* provide fascinating discussions on engineering concepts, inventions and entrepreneurship. BBC programmes can also be downloaded as podcasts and listened to via BBC Sounds, with options including *Inside Science*, *The Life Scientific* and *The Infinite Monkey Cage*. For wider perspectives on engineering advancements, platforms like Spotify, Apple Podcasts and Google Podcasts feature series such as *Engineering Matters*, *Omega Tau: Science & Engineering* and *Future of Engineering*.

Examples of your areas of interest

One way to make an interview a success is to illustrate the points you are making with examples. It is also easier to talk about something you know about rather than trying to talk in general terms. And if the examples you use are interesting, the interviewer may well want to talk about them rather than ask you the next question on his or her list. But remember, this will only work if you have done your research beforehand. There is nothing worse in an interview than a conversation along the lines of:

You: One of the things that inspired me to study civil engineering was a journey with my parents up the east coast of England when we crossed the Humber Bridge, the first suspension bridge I had ever seen at first-hand.
Interviewer: I see. Can you tell me something about the reasons for building a suspension bridge there rather than a beam bridge?
You: Sorry, I don't know.

A better answer would have been:

You: The Humber estuary is used for shipping, and because of the width of the estuary, a traditional beam or cantilever structure would not have been able to span the space between the banks. Also, suspension bridges have some flexibility, and that area is prone to high winds.

The interviewer might then have gone on to discuss the types of forces that are present in a suspension bridge and about suitable materials, all of which you would be familiar with because you had anticipated this response and had prepared for it.

Here are some ideas to use as examples to illustrate points you want to make:

- civil engineering: names and construction details of bridges or a new airport;
- structural engineering: how the use of new materials helped in the building of a new skyscraper or office block;
- mechanical engineering: examples of machines that you can discuss (cars, aircraft, wind turbines etc.);
- chemical engineering: an industrial chemical process;
- biomedical engineering: a medical breakthrough that was developed by engineers (such as computerised tomography (CT) scanners);
- electronic engineering: the background on the development of, for example, solid-state memory for computers.

The interview

Interview questions are likely to test your knowledge of engineering projects and developments in the real world, since, unlike some theoretical science subjects, engineering is a practical subject aimed at making the world a better place. It is important that your answers are delivered in appropriate language. You will impress interviewers with fluent use of precise technical terms, and thus detailed knowledge of the definitions of words and phrases used in engineering is essential. Potential electrical engineers need to know the technical and microscopic differences between semiconductors and insulators, and to be able to differentiate clearly between electric charge and current; and if you are interested in materials or civil engineering, you need to use words such as stress, strain, elasticity, strength, toughness and stiffness with their scientific, rather than their everyday, meanings.

One popular question is to ask which topics you have enjoyed studying at school. Be prepared for this by doing some revision so that you are not desperately trying to remember details from things you studied a year ago. Try to talk about something that is closely linked to engineering.

You may be asked about your future career plans. If you are applying for a particular field of engineering, then your future area of speciality will be apparent, but you may be applying for a general engineering course or one where you specialise only in the second year, so be prepared to talk about your plans. You may have ideas about where you want to work, in a big company or possibly overseas. It is a good idea to relate your possible plans to research you have done or to your work experience. It is also a good idea to demonstrate a knowledge of how you gain Chartered Engineer status.

You may be asked questions that appear to want your opinion on a recent development or issue. This type of question is asked to see whether you have been thinking about engineering issues, or whether you have been keeping up to date with current issues. Ultimately, the interviewer is not really interested in your opinion but in your ability to formulate arguments and your interest in the field.

Practice of interview situations, like most other things in life, will make you better prepared, less nervous and more confident. Arrange mock interviews with teachers, friends of your family or with careers advisers.

Some things that can adversely affect the interview include:
- arriving late and flustered;
- being unpresentable or scruffy;
- being unable to talk about things mentioned in the personal statement;
- not listening to the question carefully before answering;
- interrupting the interviewer.

Student experience

'Regarding interviews for mechanical engineering, I would recommend revising calculus, graph sketching and physics, such as mechanics and electricity. I used a site called www.i-want-to-study-engineering.org which has many questions and solutions for many engineering type interview questions. Also, I found that revising from my school textbook was useful for brushing up on old techniques. My interview experiences were as follows:

'First interview:

'I was given a situation where a hopper was filled with sand and placed on some scales. The task was to draw the graph of the reading the scales would show if you let the sand fall out of the hopper and onto the scales, from the time the first grain of sand started to fall until the last one fell. The next question was about a cube hanging from one of its vertices and the task was to find the shape you would get if you cut horizontally across it from different heights. (No drawing was allowed.)

'Second interview:

'I sat a one-hour paper before the interview and we went over it with the interviewer.

'To start, there were some graphs to draw:

$y=(x+1)/(x-1)$

$y=(x+1)^2/(x-1)$

'Then there were questions about circuits with resistors in parallel and in series. In a circuit with a cell of voltage V, there are two resistors in parallel to each other, which are then connected in series to another pair of resistors which are also in parallel to each other. The task was to find the power dissipated by one of the resistors.'

DS Barreto Romero, University College London,
MEng Mechanical Engineering with Business Finance

Fifty sample interview questions

A. About you

1. What first started your interest in engineering?
2. Why do you want to be an engineer?
3. What qualities does it take to be a successful engineer?
4. What have you done to investigate engineering as a course?
5. What have you done to investigate engineering as a career?
6. What field of engineering particularly interests you?
7. Tell me about what you learnt during your work experience and projects at [insert specific company].

B. Tell me about what you learnt during your work experience or projects at (insert specific company). Your research into engineering

8. As an engineer, you will be involved in creating commercially viable products. Can you discuss a situation where the commercial pressures are in conflict with environmental issues?
9. Give me a very brief outline of the key engineering developments of the twentieth/twenty-first century.
10. What role do you think AI or automation will play in the future of engineering?
11. What do you consider to be the most significant engineering project in history?
12. Do you have an engineering hero/heroine?
13. What do you think is the difference between science and engineering?

C. Academic questions

14. How does the structure of a metal determine its properties?
15. Why does the molecular structure of wood make it suitable for some building projects but not others?
16. What is the difference between a 'tough' material and a 'strong' material?
17. What do we mean by potential difference?
18. What is the difference between charge and current?
19. Why are the concrete girders used to construct buildings 'T'-shaped in cross-section?
20. How does the internet work?
21. What is meant by conservation of energy?
22. What is meant by a 'cantilever'? How do the stresses on a bridge using cantilevers differ from a bridge using a beam to span the distance between two supports?
23. Which parts of the electromagnetic spectrum can humans detect?
24. Can you explain what is meant by 'proof by induction'?
25. Can a scientific theory ever be proved?
26. What is a semiconductor?

27. What is a machine?
28. What is a computer?
29. What is a bit, and what is a byte?
30. What is 27 in binary? What would it be in a system based on the number 4 rather than 2? Or in a number system based on the number 9?
31. Why is nanotechnology so called?
32. How does a car engine work?
33. Why is the distance that an electric vehicle can travel so small compared with a petrol-fuelled vehicle?
34. How does an aircraft stay in the air when it is more dense than air?
35. What is meant by the words 'digital' and 'analogue' when describing communication systems such as TV signals?
36. Why does a bicycle have gears?
37. What limits the maximum height of a proposed new office development?
38. What is a robot?
39. An architect designs a 50-storey office building. How might an engineer test whether it is safe to build it?
40. Can you show the forces acting on a ladder?
41. A designer creates a model for a new type of passenger aircraft. How might an engineer test whether it is safe to build it?

D. **Unexpected questions (designed to test your ability to think creatively)**

42. Murphy's law says that whatever can go wrong will go wrong; for example, if you drop a piece of bread that has jam on one side on the floor, it will always fall with the jam side down. How would you go about verifying Murphy's law?
43. Engineers cause more problems than they solve: do you agree?
44. Is there any such thing as a 'new' invention?
45. Emerging technologies like 3D printing and AI are shaping engineering practices. Which of these developments do you find most exciting, and why? ?
46. People describe a good idea as being 'the best thing since sliced bread' – what are the advantages and disadvantages of sliced bread?.
47. What do you think the future of Formula One looks like for an engineer?
48. What will replace fossil fuels when they run out?
49. What do we mean by 'alternative' energy sources?
50. Describe this (showing the student an everyday object – a chair, a frying pan, a light bulb, a shoe, a watch) from an engineering perspective.

6| Succeeding at Interview

How to answer interview questions

Introductory questions

Why have you chosen to apply here?
The interviewer will need to be reassured that you have done your research and that you are applying to the university for the right reasons, rather than because your friend tells you that the social life at that particular university is excellent.

Your answer should, if possible, include the following points:

- first-hand knowledge of the university, for example, you came to an open day or you have spoken to students who have studied there; if you cannot visit the university, then at least try to discuss the institution with current or ex-students (many university websites have links to current students who can answer your questions directly);
- detailed knowledge of the course and why it is attractive to you, or how it links to your future career plans. The course might, for example, offer the chance to learn a language as one of the options in Year 2, and you could mention this as being something that will help you to work overseas. Or it might offer work placements or the chance to spend a period of time at an overseas university.

Why do you want to be an engineer?
(This question, or a variant on it – What have you done to investigate engineering? When did you decide that engineering was the right course for you? – will almost certainly be asked. It would be considered by the interviewers to be a gentle introduction to the interview because they will assume that you have thought about this and anticipated it being asked.)

Your answer should include an indication of how your interest started (e.g. taking apart a radio, building a model out of Lego or something you were taught in a science lesson) and lead on to things you have done to investigate engineering. This would, ideally, involve work experience or an engineering lecture you went to. You could end up talking about a particular area of interest (mechanical engineering, civil engineering) or a project that interests you (a building, a machine?) and possible plans for your future career.

General questions about engineering

What qualities should an engineer possess?
(Variants on this question might include, 'From your work experience, what did you learn about what it takes to be a successful engineer?')

Points you might raise could include mathematical ability, logic, analytical and problem-solving skills, and curiosity. But it is important to expand on these rather than simply list them. Explain, ideally using an example to illustrate what you are saying, why you think that this quality is important. Examples can be drawn from your work experience, your wider reading or a lecture.

Here is an example:

> *The ability to solve problems is very important. I really became aware of this when I was doing my work experience at a local engineering company. They were making low-voltage lamps for use in recessed lighting fittings in houses and offices, but in one particular building, the lamps kept blowing. In the end, one of the engineers decided that the problem must have been in the transformer rather than the lamp itself, and so he looked at where the transformers for each lamp were situated. It turned out that they were short-circuiting because the cavity above the false ceiling was damp.*

What does a chemical engineer do?
(Variants on this could include asking for definitions of engineering, science or technology.)

Again, try to illustrate your answer with an example or the details of a conversation that you had with an engineer:

> *Chemical engineers primarily design chemical manufacturing processes and then find development initiatives to enable those schemes. I was told on my work experience with a chemical engineering firm, that in order to be successful as a chemical engineer, you need to have a proficient knowledge of the principles of all three sciences and mathematics in order to address problems such as the production or use of chemicals, the development of fuels, food, drugs, etc. It is not about being left- or right-brained, to be creative – as you must be – you need to be using both.*

Questions designed to assess your clarity of thought

You may be given an open-ended question about something you have already studied. The point of this type of question is not so much to test your knowledge or academic level (because this will be clear in the grade predictions and exam results on the UCAS application) but to see if you can think logically and in a structured way. So, your answer should really be an exercise in 'thinking aloud'; that is, talking the interviewer through the steps to your final answer. An example is given below.

Why are metals so useful to engineers?
You could start from first principles by describing the microscopic structure of a metal. This shows that you can approach problems in a logical way while also giving you some time to think about where your

answer is going: 'A metallic structure consists of a lattice of positive ions surrounded by a "sea" of delocalised electrons. It is this structure that gives metals their useful properties.'

You might then go on to look at a number of properties in detail: 'The most obvious properties this gives metals are good electrical and thermal conductivity. Electrical conduction is through the flow of electrons through the lattice, and since they are not attached to any particular atom, they can move freely. Metals are also good conductors of heat because the electrons are able to transfer energy as they move in addition to the vibrations of the lattice.'

You could then move on to other properties that make metals so useful, describing each one in turn. You would probably include malleability and the use of physical processes to alter the strength, stiffness or toughness of a metal to suit its intended usage.

Questions that assess your ability to analyse or to solve problems

You may well be confronted by a question about a situation that you have not covered in your studies. Don't worry. The interviewer will know that this is a new area for you. What he or she is looking for is not for you to immediately give them the correct answer, but rather how you can take things that you know and apply them to new situations.

Applicants for one university were asked, 'What percentage of the world's water is in one cow?' Of course, no one (including the person who asked the question) knows the answer to this. What they were interested in, as discussed in the previous example, was in the candidate's ability to approach a problem from first principles and to arrive at an answer in a logical and structured way. So, an answer of 'I don't know' would not be very useful to your chances of being offered a place.

A better answer might start with: 'Well, I suppose I might begin with trying to estimate how much water there is on Earth. I know from my physics what the radius of the Earth is, and so I could work out its surface area. I could then make an assumption about the average depth of the oceans and work out their volume.' Rather than listen in silence, it is likely that the interviewer will help you by giving you hints or guiding you. But they can only do this if you explain every step.

When I talk to students about what they worry about when they are preparing for their interviews, they always say, 'What if I cannot answer a question?' And here is what I say to them.

- Interviewers are aware of the level you have studied to, and so will have a good idea of what you should know and may not know.

- Therefore, it is likely that any 'new' topic that you are confronted with at the interview has been asked to see how you think rather than what you know.
- Approach all questions from first principles, for example, GCSE knowledge, and then build up your answer.
- If your answer requires you to draw a sketch or do a calculation, ask if you can use a piece of paper or the whiteboard in the interview room.
- Don't be afraid to ask for help, but do this by asking for comments on what you think is the right approach: 'I think I would start by looking at the forces on the body – is this right?'

Questions to show your interest in engineering

Anyone can say that they are interested in engineering, but by applying to study engineering at university, you are embarking not just on a short period of study but on your future career as well. An interviewer (who is almost certainly an engineer) will want to be reassured that you are serious enough about the profession to keep up to date with developments and engineering issues. So, questions such as 'Tell me about an engineering issue that you have read about recently' are designed to see if you keep abreast of current events. How do you ensure that you are prepared for such questions?

- Watch the television or listen to the radio news on a daily basis. Read the quality newspapers as often as you can, and keep a scrapbook of engineering-related stories.
- Check websites for engineering stories. A good starting point is the BBC website (www.bbc.com), which has a section on science and technology.
- Visit the news sections of the websites of the engineering institutions and professional bodies (see Chapter 11 for website addresses).
- Talk to engineers.
- If you can, go to public lectures at universities. Details can be found on the university websites.
- Download or listen to radio programmes such as BBC Radio 4's *Material World*.

Steering the interview

There will be issues that you want to raise in the interview, things that will demonstrate your research, commitment and personal qualities. Rather than walking out of the interview disappointed that you did not have the opportunity to discuss these things, try to bring them into the conversation. For example, you may have been to a lecture on developments within the electronics industry at a local university one evening, and you want to talk about this. There are likely to be many ways that you can do this. You might be asked why you want to be an engineer, and during your answer you could say, 'and the thing that really convinced me that electronic engineering was the right career for me was listening to Professor Smith talking about nanotechnology at a lecture that I attended at Surrey University last month'.

In all probability, the interviewer will then ask you more about this, and you can then talk about something that you know about, rather than having to face questions on a topic with which you are less familiar.

Before you go to the interview, write down a list of things you want to talk about, and think of ways that you may be able to do so.

Current issues

As a potential university engineering student, you need to demonstrate your interest by keeping up to date with current issues and developments. Engineering is an ever-evolving subject, with new materials, processes and products being developed every day. Just think about the rapid changes in the field of communications over the past 15 years, for example. Once you have identified your particular areas of interest, you need to keep researching and reading about them, perhaps keeping a scrapbook (either a physical one or on your computer) of news articles that are relevant to your chosen area of study, and using this as part of your preparation for an interview.

In this section, I have covered some broad topic areas that should be familiar to potential engineers. The summaries in this chapter are included to give you an illustration of the kind of events and other news items you should be reading about. In other words, they are a starting point for your own research, rather than being an easily accessible source of information to memorise prior to your interviews.

Engineering and materials

Engineering is about making practical things, and so engineering projects require materials – that is obvious! But what factors determine which materials are suitable for a particular project? And how do engineers go about choosing between materials that may appear to have similar properties? To do this, engineers need to be able to describe the properties of the materials they are looking for. Physicists and engineers have to be very precise in the way they describe the properties of materials, and words that we use in a very general sense in our daily lives have very specific meanings for engineers. Words such as 'tough' and 'strong' or 'elastic' and 'plastic', words that we use in normal conversation and that might appear to mean similar things, have precise definitions when used in an engineering context. Prospective engineers need to learn the vocabulary of materials.

When we talk about properties, we mean how a particular material behaves when subjected to external factors – light, forces, heat, electricity and time. And a property that might be useful in one context, for example, stretching elastically, might be problematic in a different situation.

Some of the properties of materials that engineers are interested in are:

- electrical conductivity;
- thermal conductivity;
- how they behave when subjected to forces;
- how their behaviour changes over time;
- resistance to corrosion;
- how they are affected by heat or temperature;
- how different materials can be joined;
- resistance to chemicals or changing weather conditions;
- toxicity;
- aesthetic properties – will people want to buy or use the product?

Consider your smartphone. What materials are used and why? Electronic components rely on their ability to conduct (or not) electricity. The electronic circuits and microprocessors in your phone are built out of semiconductors. Semiconductors are somewhere in between electrical conductors and electrical insulators, and the way that they conduct electricity depends on the materials used. And so electronic engineers are interested in the properties of semiconductors. The flow of electricity through your phone through the conducting parts of the circuits generates heat, and so the engineer is also concerned with minimising this and thus needs to understand the different metals that are available. Some of these are cheap, some expensive, some last a long time, some corrode, some are safe to be in contact with and some are toxic.

Your phone relies on a battery to supply the electricity that it needs to function. This battery contains chemicals. It needs to be able to be recharged regularly without deterioration of the chemicals and to have a long lifetime. And what happens when the phone is defunct and has to be thrown away? What effect will these chemicals have on the environment? How safe are the batteries? If you were the engineer that developed the batteries in a recently available smartphone that appeared to have a tendency to catch fire and the company that developed it had to recall millions that had been sold, your future job prospects might not be that promising.

What does the designer want the phone to look like? Should it be available in different colours or look metallic? And how resistant will it be to wear and tear or accidents? Some materials are shiny and some are transparent. Some are strong; some will crack when the phone is dropped. Some get hot very quickly. Materials engineering is a field that answers questions like these.

Let's look at another example: building a new iconic skyscraper. An architect will create the initial design of the building, which will be based on aesthetic ideas, as well as thinking about what the building will be used for – 'form and function'. This would probably start as a rough sketch, and then the architectural practice will work on more detailed designs. This is where the structural engineers will be involved. First and foremost, the building has to be structurally viable; that is, it shouldn't fall down. So the structural engineer will need to look at all aspects of the structure: will it be built around a central concrete core with the exterior fixed to this, or are the external walls going to be load-bearing? Concrete is very strong when subjected to compressive forces but cracks easily when tensile (stretching) forces are applied. So if tensile forces are involved, then it has to be strengthened or toughened (see below) by adding other materials to reinforce it or pre-stress it. If the building is going to be tall, then how flexible does it need to be in strong winds? How will people get to the upper floors? What materials should be used for the lift cables? What about ventilation? What type of glass? A building with a glass exterior will get very hot because of the greenhouse effect.

A brief explanation of some of the commonly used terms that describe materials or their properties is given below. But you should read more about the properties of the materials that are used in the particular field of engineering that interests you.

Types of materials
Metals: the atomic structure of a metal is what gives metals their most useful properties. Because the electrons are relatively free to move through the fixed crystal lattice structure, metals are good conductors of electricity and heat. These 'delocalised' electrons also make metals malleable and shiny.

Polymers: made up of long chains of repeating units. These molecular chains can be arranged in more or less orderly patterns or tangled up randomly. This means that polymers can usually be stretched and can be moulded into shapes. Polymers are chosen for particular uses by looking at their elastic or plastic behaviour and by looking at how they react to heat, light or chemical exposure. In October 2018, the introduction of self-healing materials on an industrial scale moved ever closer. Chemical engineers from the Massachusetts Institute of Technology (MIT) designed a polymer that is able to react with CO_2 from the air, which would then grow, strengthen and potentially even be able to repair itself. It is designed as a way of positively using greenhouse gases, and researchers claim that it could, in the future, be used for construction material or a protective base layer.

Amorphous: as the name suggests, the atoms and molecules are randomly arranged with no long-range discernible order. Glass and ceramic materials are amorphous in structure.

Composites: composites are composed of different materials (or the same material in different orientations) and exhibit the best properties of each. Reinforced concrete incorporates steel rods within the structure, so it is strong in compression and also in tension.

Properties of materials

Strength: strength is a measure of how much stress a material can withstand (a similar measure to pressure) before breaking.

Toughness: not the same as strength. A tough material might or might not be strong, but it will not be brittle; that is, it won't break through cracks appearing and spreading through the material. So glass can be strong, but is not tough (unless it is laminated or treated to improve its toughness), while a digestive biscuit is neither strong nor tough. A polythene bag might not be very strong but it is tough.

Stiffness: a measure of how easily the material deforms – bends, stretches or compresses – when forces are applied.

Brittleness: brittle materials tend not to stretch very much when a force is applied but then suddenly break. Glass is a brittle material.

Elasticity: elastic materials deform when forces are applied but then return to their original shape and size when the force is removed. Why is an elastic band so named? Because it exhibits elastic behaviour.

Plastic behaviour: the opposite of elastic. If the force is removed, the material does not return to its original shape. Take a plastic supermarket bag and cut a thin strip from it and then stretch it. Initially it will not stretch much, but when you apply a certain force, it will suddenly be easy to stretch by a significant amount. But when you stop applying the force, it will not go back to its original size. So a plastic bag is so called because it behaves plastically.

Malleability: metals are malleable. It means they can be hammered or pressed into new shapes.

Electrical conductivity: when a potential difference is applied to a conductor, how much current will flow? The higher the current, the higher the conductivity. Electrical conductivity may be a desirable or undesirable property. Electrical insulators are also required in any devices that use electricity. Semiconductors are materials that conduct electricity by an amount determined by how they are made or at what temperature they operate.

Thermal conductivity: a measure of how much heat flows in a material. As with electrical conductivity, both good and poor thermal conductors are useful to engineers. Cooling is important in many mechanical devices where heat is generated, and so good conductors are needed to remove the heat efficiently. But the handle of your kettle should be a good thermal insulator so you do not burn your hand.

Failure mechanisms: a material or component can fracture or stop being suitable for its chosen purpose in many different ways and from many different causes. They can crack, deform, melt, corrode, decay, suffer from metal fatigue or stop being useful in many other ways. The causes of this could be defects in the molecular structure (metal fatigue is caused by dislocations within the structure), the application of too much force or heat or electric current, through chemical reactions, surface scratches or even by being joined to other materials using unsuitable adhesives or techniques.

Commercial issues

Someone has to pay for the things that engineers produce. Sometimes you pay through your taxes (roads, bridges, medical equipment), but in most cases, the things that engineers make have to be useful or attractive enough for people or companies to pay for them. Two very similar, in terms of functionality, smartphones or cars may vary in price by a factor of four or five simply because of the materials used.

Engineering, ethics and the environment

Engineers are professionals, and, as such, they have to behave in a professional manner and ensure that they act ethically at all times. Engineers need to be trusted by the public. The work that engineers do affects millions of people – think of engineering projects that involve defence, power generation or the provision of clean water, for example.

The Royal Academy of Engineering and the Engineering Council sum up the obligations of engineers in their Statement of Ethical Principles:

1. Honesty and integrity;
2. Respect for life, law, the environment and public good;
3. Accuracy and rigour;
4. Leadership and communication.

Source: www.engc.org.uk/media/2334/ ethical-statement-2017.pdf.

In the publication 'Engineering Ethics in Practice' (available to download at https://raeng.org.uk/ethics), there are case studies that outline how these principles apply to real-life situations. They illustrate how, for example, engineers need to get the right balance between the economic demands of the companies that they work for and ethics. An example of this is the ethical concerns around AI and machine learning, particularly in sectors like healthcare, criminal justice and hiring. In 2023, several companies faced backlash over the biased outcomes produced by AI systems, which raised ethical questions about fairness, accountability and transparency. Additionally, there has been growing scrutiny over the energy consumption of AI systems themselves, with calls for more energy-efficient algorithms and AI models to reduce the environmental footprint of data centres.

Other ethical issues involve surveillance, cutting costs to ensure that projects finish on budget or the use of poor-quality or dangerous materials.

The environment

Almost everything that engineers produce, design or operate has a potential impact on the environment; they are at the forefront of the global climate crisis, and it will be they – alongside the collective will – that will resolve it. Some large-scale projects have an obvious effect – a new dam, wind farms, new housing or office developments. Cars and aeroplanes produce emissions as a result of converting fuel into kinetic energy. Power stations produce electricity from fossil fuels or nuclear reactions, and the by-products of these can be harmful to living things or have long-term effects on climate. Discarded batteries and electronic devices contain poisonous chemicals. Engineers are increasingly having to incorporate safeguards into their designs as public pressure is put on them to reduce environmental damage and avoid the depletion of natural resources and habitats.

Green buildings

It is estimated that between 30% and 40% of energy use in developed countries is associated with buildings, most of which is used in heating, cooling and electrical systems. The concept of a 'green building' is a relatively new one. By describing a building as being 'green', we mean that it is designed to be as energy efficient as possible. This can be achieved by:

- using building materials that do not require large amounts of energy to produce – plastics, concrete and metals require significant amounts of energy, whereas natural materials such as wood require much less;
- insulating buildings effectively to avoid heat losses;

- designing buildings with natural ventilation, such as using natural convection to produce cooling, rather than relying on air conditioning;
- aligning buildings so that sunlight can be used either as a source of heating or of electricity in photovoltaic cells;
- using 'thermal mass' to absorb heat and then release it naturally;
- re-using water through recycling facilities.

Aerospace

Extreme weather in space has now been upgraded to the National Risk Assessment (NRA). The Royal Academy of Engineering is in consultation with relevant partners over the establishment of infrastructure that is able to withstand solar storms. For example, Global Navigation Satellite Systems (GNSS) and developing GNSS's capacity to operate with holdover technology for up to three days. Extreme space weather is a clear and present concern, and this is likely to be the focus of certain strands of engineering from now on.

In 2022, NASA launched its Psyche mission into deep space in order to experiment on the future of deep space communications using lasers. And of course, commercial space flight is now eminently possible and has been achieved, though the price bracket is still unachievable for the 99%. Still, as a feat of design and aeronautical engineering, it is extraordinary.

Hybrid vehicles

Hybrid vehicles are vehicles – cars, trains and trams – that have two separate sources of energy. While the idea is not new (the moped is a hybrid vehicle as it is powered by a motor or engine and by the rider), car manufacturers are now devoting enormous resources to new ranges of hybrid electric vehicles. Hybrid petroleum/electric vehicles are powered by internal combustion engines that run on petrol, diesel or gas. They also contain heavy-duty batteries that are charged as the engine is running, and the car can be powered electrically during parts of its journey. Other hybrid vehicles use the engine to compress air or hydraulic fluid, which can then be used to drive the wheels, and there are trams that are able to use a combination of diesel fuel and overhead electric cables. What all of these have in common is that they reduce the use of fossil fuels (and so reduce greenhouse gases and pollution) and utilise energy that might otherwise be wasted when the car is running but not moving.

Other examples of how engineers are adapting to the needs of the environment include:

- the development of new energy-efficient materials, such as composites that make aircraft bodies lighter or insulators that prevent heat losses;

- electrical conductors that have low electrical resistance in order to reduce heating when conducting electricity;
- stronger, lighter materials for electrical devices;
- ways of extracting minerals or fuels from below the Earth's surface that have lower environmental impact;
- new ways of producing energy that do not rely on fossil fuels (see below).

Sustainable transportation infrastructure

Sustainable transportation infrastructure is a rapidly developing field that aims to reduce carbon emissions and improve urban mobility. One of the most exciting innovations is the development of electric vehicle charging networks, which are being integrated into urban planning to support the growing adoption of electric vehicles. Governments and private companies are deploying fast-charging stations that enable electric vehicle owners to quickly charge their vehicles on the go, reducing range anxiety and encouraging greater EV adoption.

Intelligent transport systems that use AI are becoming increasingly prevalent. These systems can optimise traffic flow, reduce congestion and improve the efficiency of public transport systems. The development of 'smart cities' with interconnected transportation networks is enhancing mobility, energy efficiency and sustainability in urban areas.

Plastic waste

The ever-increasing amount of plastic in the world's oceans is a clear and present problem. Engineers are looking at ways of combating this issue. Innovations such as Seabins – which drag in, on average, over 4 kilograms of debris per day – sea walls and water wheels in rivers to catch debris have all been trialled. The key is a global, unified effort: engineers everywhere need to work as a team to combat a problem that continues to escalate. Organisations such as Ocean CleanUp are at the forefront of this, and you should research them before your interview.

Engineering and energy

Engineers use energy to create their products and to run them. Engineers also create the structures and processes that produce usable energy. Engineers, therefore, are at the forefront of the search to develop new energy sources. As well as being finite sources of energy (in other words, they will one day run out, or, at the very least, become uneconomical to extract), fossil fuels cause environmental damage and are closely linked to climate change and global warming. Many fields of engineering have an interest in developing alternative or renewable energy sources. While there are no set definitions of these words, they generally refer to energy sources that do not involve fossil fuels and/or those whose consumption does not deplete the planet's natural resources or reserves. The engineers who are primarily involved in developing alternative energy sources are electrical engineers, electronic engineers, bioengineers, mechanical engineers, geoengineers and chemical engineers.

There are a number of ways of categorising energy sources (see Tables 2 and 3).

Table 2 Categorisation of energy sources based on origin

Energy sources derived from the sun	Energy sources not derived from the sun
Solar	Geothermal
Wind (heat from the sun creates areas of low and high pressure, causing wind)	Nuclear fusion
Wave (winds cause waves in the sea)	Nuclear fission
Fossil fuels (coal, oil, gas and so on were once living things)	Tidal
Hydroelectric (water sources, such as rivers, were created by precipitation of rain, caused by the sun's heating)	
Biomass	

Table 3 Categorisation of energy sources based on effect on source

Renewable*	Non-renewable
Solar	Fossil fuels
Wind	Nuclear fission
Wave	
Tidal	
Hydroelectric	
Geothermal	
Biomass (since although the individual sources cannot be used again once the energy is extracted, the supply can be maintained by, for example, planting more oil palms)	
Nuclear fusion (considered renewable since the likely source would be sea water, which is effectively inexhaustible)	

*By 'renewable', we generally mean that the source of the energy is unaffected by the extraction of energy. For example, every time coal is burned there is less coal remaining on the Earth, whereas if a wind turbine creates electricity, it does not affect the source (movement of air, which is caused by solar heating).

Alternative energy sources that are currently used commercially are:

- geothermal energy;
- hydroelectric power;
- nuclear fission;
- green hydrogen;
- solar power (solar panels to produce heat and photovoltaic cells that create electricity from sunlight);
- tidal power;
- biomass (using bacteria to produce gas through the digestion of organic materials, making fuels out of natural materials such as palm oil);
- wave power;
- wind power.

An alternative energy source that is in development is nuclear fusion.

Recently, fracking has become one of the most contentious new ways of sourcing fuel, causing split opinion across communities. It is worthwhile understanding the arguments for and against this process while considering it from an engineer's perspective, that is, the hydraulic process involved in splitting the rocks and the repercussions of doing so.

Engineering and the NHS

Right now, the NHS needs innovation. The only way it can continue to develop and aid the lives of others is to have the technology to support it. With the day-to-day pressures faced by the NHS, the engineering profession is finding new ways to support the needs of the health service for the benefit of those working in it and those using it. Much like the healthcare system, engineering is systems-based and is a sector always looking for innovation. It has long been acknowledged by those within both sectors that cross-disciplinary learning could yield new understanding and greater efficiency.

Engineering in the NHS could mean anything from the regeneration of ageing hospitals to inventing new medical technology. In recent years, the integration of digital health technologies has become a key area of focus, with engineers developing solutions for telemedicine, wearable health devices and AI-driven diagnostic tools. Biomedical Engineering is now an established degree course.

In October 2018, King's College London announced that it was teaming up with engineering firms Nvidia and their AI Tech to utilise their supercomputer in the fields of radiology and pathology. Given that the number of patients requiring radiology is reportedly increasing around 16% every year yet the number of radiologists required to analyse that data is nowhere near in step, the supercomputer could provide a much faster analysis of imagery. More recently, AI algorithms are being developed to assist in early diagnosis of conditions like cancer and heart disease, with systems that can analyse medical images faster and with greater accuracy than human doctors. This is just one example of engineering partnerships through technology that can aid the NHS in the long term, but given that medicine is a technology-focused world, you should be able to find many more examples before your interview.

Remember, too, that when we talk about engineering within the medical field, it does not have to be technology-related; for example, the development of 3D-printed prosthetics is one of the most progressive fields of biomedical engineering.

The commercial space race

Elon Musk (Tesla and SpaceX), Jeff Bezos (Blue Origin) or Sir Richard Branson (Virgin Galactic)? Currently, Forbes has the US ahead of Europe in the race to send a commercial flight into space. In 2018, Musk's Tesla sent a car into space on a privately made rocket. In 2020, Musk's SpaceX announced that it had developed the most powerful rocket ever: Falcon Heavy; Elon Musk also has the grand ambition of flying to Mars. Well, Richard Branson and then Jeff Bezos achieved their ambition of going to space in 2021, while Elon Musk is still yet to make it. However, SpaceX is the only commercial space company with the capability of sending astronauts into orbit (the others are suborbital; technically space). What are your views on that? Anything is possible; it is just a question of when and how commonplace it will become. Commercialisation is a term often used in this conversation, but remember that due to a price tag of at least US$250,000 per seat, we will not be looking at an overcrowded space channel for at least the immediate future.

Aviation and aeronautical engineers are at the forefront of pioneering these passenger spacecraft. Their design and structure are critical, as there are different factors to consider when operating for a commercial audience, and a great variety of engineering disciplines will be utilised in the process.

Sport and engineering

The combination of sport and engineering could take many forms, from training enhancement to recovery equipment. It could also reference infrastructure and the vast array of building projects for the major sporting teams, from the redevelopment of the pre-existing Anfield site to the widescale complete build of several major stadiums for the 2022 FIFA World Cup in Qatar. Try and stay neutral in your political views, and keep your comments specific to what the degree is actually about.

New cities?

Is there space left on this planet for new and sizeable cities? Yes, and they are in the desert. Egypt has long been developing extra city space, following on from Dubai and Abu Dhabi. However, by far the most iconic and eye-catching at the moment is in Saudi Arabia with the construction of NEOM. NEOM plans to be the world's first city built in a single straight line, measuring 110 miles long and 200 metres wide. It is to be 100% run on renewable energy, and 95% of the land will be reserved for nature. What are your views on such a wide-scale building project?

Artificial Intelligence and automation

The use of AI and automation in engineering is rapidly transforming industries and processes, leading to advancements in design, production and maintenance. AI algorithms are used in predictive maintenance systems, ensuring that everything from machinery in factories to aircraft engines operates efficiently by predicting failures before they occur. Automation has revolutionised manufacturing, with robotics enhancing production lines and speeding up processes while heightening accuracy and reducing waste.

The construction industry is another area being reshaped, with automated drones and robotic bricklayers redefining traditional building methods, increasing speed and safety. Meanwhile, digital tools such as generative design software utilise AI to create optimised structures based on specified parameters, exploring thousands of design possibilities within moments – an approach already applied in industries like aerospace and automotive manufacturing.

Looking to the future, autonomous vehicles are perhaps the most high-profile example of engineering innovation supported by AI, from self-driving cars to unmanned aircraft systems. These projects require a blend of mechanical, electrical and software engineering skills to solve challenges in safety, navigation and infrastructure readiness. How do you think AI and automation will balance progress with job displacement in engineering sectors, and how might these technologies further improve sustainability and efficiency in the future?

Engineering business case histories

Try to keep up to date with stories associated with engineering businesses and entrepreneurs. Most new engineering projects and developments are driven by commercial companies that want to sell their products or processes, and while some new devices are developed by university research departments, these are often funded by industry.

6| Succeeding at Interview

The following list aims to provide you with a starting point for your research, as there will undoubtedly be new names that will become more relevant by the time you read this book:

- Apple's iPad and iPhone and Apple Watch, the Android, iOS and Windows operating systems for mobile devices;
- Google;
- Apps;
- Amazon and its innovation in logistics and automation;
- NEOM;
- Virtual Reality (VR) and Augmented Reality (AR);
- Tesla's electric vehicles;
- recyclable waste and green energy;
- Formula E;
- HS2;
- 5G networks;
- inventors, scientists and entrepreneurs, such as Elon Musk or Tim Berners-Lee;
- SpaceX or SpaceShipTwo;
- Virgin Galactic;
- multinational construction companies, such as Arup;
- aeronautical engineering companies, such as BAE Systems;
- 3D-printed buildings and construction;
- UK tech companies (e.g. those based in 'Silicon Fen', such as ARM);
- biotech companies, such as Gilead Sciences;
- oil and petroleum companies that are developing alternative or renewable sources of energy, such as Statoil;
- online school platforms as educational engineering to deliver innovative education;
- 4D cinematic technology;
- digital money;
- hyper-personalised drugs;
- alternative energy schemes, such as the proposed tidal lagoon in Swansea Bay;
- building tall structures, such as the Burj Khalifa in Dubai or the Shard in London;
- Battersea Power Station and the change of ownership;
- issues surrounding the excavation and distribution of rare earth metals;
- collapse of Carillion and the knock-on effect on engineering firms such as Vaughan Engineering;
- fuel cells that produce electricity from hydrogen;
- self-heating polymers;

- how engineers deal with natural disasters, such as floods, earthquakes and tsunamis;
- environmental issues, such as the safe processing and storage of used fuel rods from nuclear power stations or polluting metals used in mobile phones or computers.

> **Did you know?**
>
> The word 'engineer' comes from the Latin *'ingenium'*, meaning cleverness.

7 | Don't get your wires crossed: Non-standard applications

Perhaps you are studying for a mixture of examination qualifications, or you have had a gap in your education. You may have already started a degree course in another discipline and want to change direction. This chapter applies to students who may be applying to university as mature students, perhaps with qualifications other than A levels or the equivalent, to international students who are applying from outside the UK and to those with disabilities.

Whatever your situation, the first thing you should do is make contact with some universities (either by telephone or via the email addresses given on the university websites) to explain your situation and ask for advice.

Changing direction

Not everyone at the age of 17 has a clear idea of where their future lies. Often, university or career choices are made on a whim – on the advice of parents, teachers or peers – or simply through lack of experience in the chosen discipline. People change, and luckily universities understand this and are used to dealing with it. Making the wrong university choice should not lock you into a study path or career that is unsuitable, unviable or unpleasant.

> **Case study: Student experience**
>
> 'I had always been interested in engineering, particularly due to my strengths in Mathematics and Physics during A levels. These subjects not only came naturally to me but also sparked a fascination with understanding how theoretical principles could be applied to solve real-world problems. As I progressed through sixth form, it became clear that pursuing Mechanical Engineering at university was the right path for me. The degree offered a perfect blend of my academic strengths and interests, along with excellent career prospects in fields like automotive design, aerospace and renewable energy.

'The application process for university felt like a natural step, and once I began my Mechanical Engineering course, I quickly realised how well it aligned with my skills and aspirations. The course structure provided a strong foundation in core engineering principles while also allowing for exploration of specialised areas that matched my career goals. From fluid dynamics to materials science, every aspect reinforced my decision to choose this degree.

'The course provided excellent support from the start. Each student is assigned a personal tutor who is available to guide and support them throughout their academic journey. My tutor was particularly helpful in offering advice on module selections and ensuring I was aware of the opportunities available within the programme, including internships and project-based learning that would help me build practical experience.

'Choosing Mechanical Engineering has not only allowed me to thrive academically but also positioned me for a rewarding career in a dynamic and ever-evolving industry. Looking back, I'm confident that my decision was the right one, driven by my passion for problem-solving, my strengths in Mathematics and Physics, and the exciting opportunities this field offers.'

Daniel Noon, MEng Mechanical Engineering

Mature students

Mature students (defined as students who are over 21 at the start of their proposed courses) are usually:

- applying with appropriate qualifications, for example, A levels, but have not been to university and are now applying after a gap of a few years;
- applying for a second degree, having graduated in a different subject; or
- applying with no A levels or equivalent qualifications.

If you come under either the first or second of these categories, you apply using the same route as first-time applicants. However, you should contact the universities directly to discuss your situation with them and to get their advice. The structure of your personal statement will need to be different from that produced by a student who is still at school or college. It should include:

- a brief summary of why you are applying now and what you have been doing since you completed school or college education;
- your reasons for the change in direction;
- an explanation of any gaps in your education or work history;
- a discussion of any appropriate skills or experiences gained in your previous jobs or degree studies.

7 | Non-standard Applications

Most universities encourage mature students who want to apply for entry to degree courses, taking into account their work experience and commitment as part of the entry criteria. Mature students have often left school without the appropriate academic qualifications for university entrance in order to start jobs or careers, in which case there are many Access courses in colleges around the country that specifically prepare mature students for higher education. Others may have studied at degree level in another, non-related, discipline.

Universities often encourage mature students to apply because they:

- bring valuable real-life experience to the faculty;
- can be more mature in their study methods than school-leavers;
- have had more time to think about what they really want to study;
- have a better understanding of the links between study and work.

The best way to find out about acceptable Access courses is to contact the universities directly and ask which ones they recognise or recommend. The engineering institutes (see Chapter 11 for a list of these) provide details of schemes that allow students who have studied on apprenticeship programmes to progress onto a Higher National Diploma or degree course.

If you are applying for a degree course as a mature student without using the Access course route, you should:

- use the UCAS application system as described earlier in this book;
- select 'no' to the question 'are you applying through your school or college' when prompted on the online application;
- fill in the section on employment as comprehensively as possible, ensuring that there are no periods of time since you left school that are unaccounted for;
- ask someone suitable (your current employer or a previous teacher) to act as your UCAS referee);
- ensure that he or she knows what the universities require from the referee (you can point them towards the UCAS website, which has a section on information for referees);
- send a more detailed CV directly to the universities once you have received your UCAS number (and quote this on all correspondence with the universities).

It is important to emphasise that the universities are keen to recruit serious, motivated and committed students onto engineering programmes, and mature students tend to fit this description extremely well. You will find that the university engineering departments are eager to help suitable candidates apply, and they will be able to provide you with advice and feedback if you contact them prior to applying.

International students

In the 2023 application cycle, UCAS (www.ucas.com) applications and acceptance statistics showed that 830 EU students (a drop from the 2022 figure of 890) and 5,925 non-EU international students gained places on engineering courses.

International students are usually:

- following A level (or equivalent) programmes either in the UK or in their home country;
- studying for local qualifications that are recognised as being equivalent to A levels in their own country; or
- studying on academic programmes that are not equivalent to A levels.

Students following A level or equivalent programmes should apply through UCAS in the usual way. All of the information in this book is equally applicable to them.

Students studying qualifications that are accepted in place of A levels can also apply through UCAS in the normal way, from their own country. The UCAS website (www.ucas.com) contains information on the equivalence of non-UK qualifications. These include the Irish Leaving Certificate, European Baccalaureate and some international O levels and A levels. Information on the equivalence of other qualifications can be found on the UK government's qualifications website (www.enic.org.uk).

Students who do not have UK-recognised qualifications will need to follow a pre-university course before applying for the degree course. These include:

- university foundation courses that are offered by many UK colleges and universities. For example, the NCUK (www.ncuk.ac.uk) programme includes specialist science and engineering foundation paths. Typically, these programmes last nine months and involve 20–25 hours of intensive tuition a week, plus specialist English help;
- university foundation courses set up by, or approved by, UK universities or colleges but taught in the students' home countries;
- A level courses (normally two years but in some cases, can be condensed into one year) in schools and colleges in the UK. A levels allow students to apply to all UK universities, including the top-ranked universities such as Oxford, Cambridge and Imperial College.

Foundation (or 'pathway') courses are not accepted by all UK universities. You should check with your preferred universities which courses they accept before committing yourself. Representatives of

UK universities, schools and colleges regularly visit many countries around the world to promote their institutions and to give advice. You can also contact the British Council to get help with your application.

International students should note that they are required to pay higher fees than UK students. UK students have their fees capped at a maximum of £9,535 per year, whereas fees for international students on engineering courses are likely to be between £10,000 and £30,000 per year. Accommodation and meals will be extra. How much living costs are depends on where you study, but, as a rough guide, around £2,000 a month should cover food, accommodation, books and some entertainment costs depending on which city you are living in.

Following Britain's departure from the EU in January 2021, EU students are no longer classed as 'Home' students and are not eligible to apply for student funding.

One of the reasons why international students are, on the face of it, less successful in their applications than UK students is that they, their teachers or their referees are unaware of what is required, particularly if they have experience with applications for universities in other countries. They have also been up against a high volume of EU students, though this may no longer be the case, following the removal of EU students' 'Home' status after Brexit. All of the information on personal statements and interviews in this book applies equally to UK and international students.

If you are applying from outside the UK, you must ensure that your referee is familiar with the latest UCAS reference requirements. UCAS has updated its reference process to be more digital and structured, with specific sections to be addressed. You can find detailed guidance for international referees on the UCAS website (www.ucas.com), which includes advice on how to provide references that align with UK university expectations, including understanding UK grading systems and qualifications. The reference should focus on key aspects such as your academic abilities, personal qualities and suitability for the course you're applying for.

English language requirements

Students whose first language is not English will be required to demonstrate that they have the right level of English language to cope with the course. Each university will specify a minimum score on an approved SELT (Secure English Language Test) such as IELTS. Students who have sat IB or IGCSE may not be required to sit an extra English language test. The UCAS Guide for International Students has details of English language requirements: www.ucas.com/international/international-students.

Checklist for international applicants

1. Check that your academic qualifications are accepted by universities. The UCAS Tariff lists all international qualifications that, in theory, are accepted. If your qualifications are not included in the Tariff, contact the universities directly.
2. Check that the level of your qualifications (grades, marks, predicted grades or predicted scores) is at the level required by the universities. The Course Search facility on the UCAS website will give you links to the universities' grade/score requirements.
3. Go to the international pages of the university websites to see if there are any specific language requirements, for example, a minimum IELTS score. Some universities will specify an overall score only, for example, 7.0. Others may also have particular requirements for each section of the test, such as 7.0 overall with at least 6.5 in each section.
4. Use the international pages to see whether there are any scholarships available to international students from your country.
5. Follow the advice in the engineering sections of the universities' websites regarding any specific information they want to see included in the personal statement.
6. Use the university websites to check the fee and accommodation arrangements.
7. If you are studying at an international school that has sent students to UK universities in previous years, it will be aware of its role in the UCAS application. The school is likely to already be registered with UCAS. If you are studying at a local school that has not sent students to UK universities, you will need to register as a private candidate on the UCAS website and discuss the reference with someone suitably qualified to write it, perhaps one of your teachers or a previous employer.
8. Be clear about the deadlines for applications.
9. Check whether you need a visa to study in the UK (see overleaf).

Visa requirements

Students from non-UK countries generally require a visa to study at degree level in the UK, although students from some countries may be eligible for streamlined application processes. The UK government website has full details of whether you will need a visa (www.gov.uk/check-uk-visa). The stages in obtaining a visa are:

1. You have been made an offer from an institution that is able to act as a licensed sponsor for your visa application (www.gov.uk/government/publications/register-of-licensed-sponsors-students).
2. You have accepted the offer and have fulfilled the academic, English language and financial requirements.

3. The institution has given you a Confirmation of Acceptance for Studies (CAS).
4. You apply for the visa, presenting the required evidence. Academic evidence, such as an IELTS certificate or examination results, will be listed on the CAS, and the gov.uk website will list other evidence, such as proof of funds to cover your tuition and living costs. Financial requirements will depend on the length of your course and its location (e.g. London or outside London).

Note: If you are from the EU, EEA, Liechtenstein or Switzerland and you enter the UK for study, you will require a student visa to attend a university in the UK.

Students with disabilities and special educational needs

Universities welcome applications from students who have physical or other disabilities or special educational needs, and they have well-established support systems in place to provide assistance and special facilities. The services offered can help with a wide range of disabilities, including sensory (visual/hearing) impairment, mental health difficulties, mobility impairment, dexterity impairment, Asperger's syndrome or other autism spectrum disorders, chronic medical conditions (e.g. diabetes, epilepsy or asthma) and specific learning difficulties (e.g. dyslexia or dyspraxia). In all cases, you should contact the universities directly, before you apply, to explain your particular needs and requirements. They will then be able to give you information on how they can help you. The website www.disabilityrightsuk.org contains useful links and information.

Case study

Duyanh went to the University of Cambridge to study Chemical Engineering. He has now started his own company in the chemicals sector.

'On completion of my A levels, I chose to study Chemical Engineering as it was closer allied to my future ambitions in my home country in Vietnam, and this is where I have been since graduating, running my own chemicals company.

'As an international student, I found the application process difficult and was grateful for the support I received. As chances for international students are fewer than for home students, I would advise anyone in a similar situation to get as much help as they can. Ultimately, what works is to make the application personal to you and not to use anyone else's words. I like to think that I got to where I wanted to be as a result of my interest in and engagement with the subject, which I was able to display in interview. You need to be strategic about where you want to apply: don't be fixated on London, for example, where application numbers are higher. You need also to remember that if you are studying A levels, unlike overseas qualifications, you are learning skills necessary for university and it is worthwhile highlighting those skills in your statement.

'Also, remember what makes you unique.

'In my first year, I took the engineering route which was the correct path for me as it gave me a broad overview of engineering. In my second year, I became particularly interested in the module on Process Operations and, specifically, Homogeneous Reactors, which made me want to do further research. This is what I enjoyed most about my course: the freedom to investigate whatever we wished, and to create and to innovate.

'I am now running my own company. The skills that an engineering degree gave me are not just applicable to the scientific value of the field, they are also relevant to how to run a company. It is a systems-based degree after all, and the modelling of a company in that way allows for a greater understanding of how to make an engineering and distribution-based company efficient and productive.

'Engineering is one of the most diverse and rewarding careers that I could have hoped to study, and even though it is arguably harder for an international student, that should just make you want it more and therefore you will work harder to achieve your ambition.'

8 | Elbow grease: Results day

You have done all of the hard work – the personal statement, the interview, the examinations – and you are now waiting for your results, the results that will determine whether you have achieved what you need to take your university place. This chapter explains what happens when you get your results and, if you have achieved grades that are either better or worse than expected, what other options are available to you.

When the results are available

- A levels – mid-August
- IB – early in July
- Scottish Highers – first week of August

Ask your school or college for the exact date and time that they will issue you the results. Whichever of the exam systems you are sitting, you need to act quickly if you:

- have missed the grades or scores that you require to satisfy your firm offer;
- are not holding any offers and wish to apply through UCAS Clearing.

What to do if you have no offers: UCAS Extra

If you apply for five courses and either receive no offers or decline all the offers you get, you are eligible for UCAS Extra. Extra operates from the end of February to the beginning of July and allows you to add one additional choice at a time.

To find a course using Extra, use the UCAS search tool and the filter 'Show courses with vacancies'. Next, contact the universities and colleges listed to check if they'll consider you. It's recommended that you call the university to which you want to apply before you add the

Extra choice to check whether there is space on the course and to discuss your suitability. To apply for the new course, you need to add the details to your application.

Your chosen university will consider your application, and if this is unsuccessful, you can add another Extra choice as long as it's before July. If you have not heard back from the university within 21 days, you can add another Extra choice (again, before July).

Once you have received an offer through Extra, you'll need to either accept or decline it. Ensure that you respond by the date displayed on your homepage, or your offer will be automatically declined.

Don't worry if you don't receive the offer you'd hoped for in UCAS Extra – you can still participate in Clearing.

What to do if things go wrong during the exams

Occasionally, students will underperform in an examination through no fault of their own. This could be due to distressing family circumstances (a serious illness of a family member, e.g.), illness in the run-up to the exam (or during the exam) or unforeseen circumstances such as late arrival to the exam due to problems with public transport. In all cases, you should inform the universities that this has happened to you immediately after the examination. You should, if possible, get your referee to give the details to the universities and provide documentary evidence, such as a letter from your GP.

What to do on results day

You can collect your results from your school or college, or you can arrange to receive them via email or post. It's a good idea to go into school or college to receive them in person so that you can get support and advice from teachers and career advisers about your options if you need it.

UCAS receives your exam results directly and will update UCAS Hub with the outcome of your university applications on results day. The system will be busy, so you might need to be patient to find out whether you've been successful.

You'll need to have the following things ready to ensure that you can do everything you might need to do on results day:

- UCAS Hub login details;
- UCAS ID number;
- UCAS Clearing number, if you go into Clearing;

- details of your offers;
- the UCAS and Clearing numbers of your chosen universities;
- a working phone and computer, so you can communicate by phone or email.

When you do get your results, one of four things will happen:

1. You receive confirmation of your place from the university you selected as your firm choice and accept it.
2. You have not met the offer from your firm choice, but you will receive confirmation from the university you selected as your insurance choice and accept it.
3. You have met and exceeded the offer made by your firm choice and decide to try to swap courses by going through Clearing (see below).
4. You have not met the requirements of any offers and need to go through Clearing.

If you have achieved the grades that meet the offer made by the university you selected as either your firm choice or insurance choice and are happy with this offer, then congratulations! You do not need to do anything. However, if you want to make use of UCAS Clearing or have not met any offers and need to use Clearing, then read on.

What to do if you exceeded the grades that you expected

If you have met and exceeded the conditional requirements of your firm choice and it has accepted you – therefore converting the conditional offer into an unconditional one – you could potentially swap your place for one on another course that you prefer. The phrase 'met and exceeded' means that if you needed BBB, you would have achieved ABB or better. It doesn't necessarily mean that you just got more UCAS points. For example, if you needed BBB and achieved A*BC, then you would have accumulated more UCAS points with A*BC than you would have if you had only achieved BBB. However, you would have still failed to meet one of your grade requirements. In cases like this, your eligibility will depend on whether your offer was based on UCAS points or grades.

If you decide to pursue a different course, you have to go through UCAS Clearing. Since more than 50,000 students get a course through Clearing, it is highly recommended that you find the course of your preference as soon as possible, as this is a first-come, first-served system.

Use the Clearing search tool to find all the available courses. Once you have found an alternative course, you will need to phone the university yourself. When you call the university, you will need to give them your UCAS personal ID number and explain straight away that you have exceeded the grades of your offers and are applying through Clearing. Be prepared to answer questions about why you really want to study that course. If they agree to accept you, and you in turn agree to accept them, this will happen during the phone call. Once you receive an offer, you can add it to your application so the college or university can officially accept you. At this stage, your status on UCAS Hub will change. Remember that if you do not find an alternative course that you want, or do not get accepted onto an alternative course, your original firm offer will still stand.

Make sure that you think carefully about the courses and universities if you decide to go through Clearing. Just because a university has higher entry requirements or is considered to be more prestigious, it does not necessarily mean that you will enjoy the course more. Consider carefully why you selected your initial firm choice and check whether your reasons are still valid and if you have the same interest and passion to study a new course.

What to do if you have no confirmed offers

If you are not holding any offers, there could be several explanations.

- You may have missed the required grades for both your firm and insurance offers.
- You may have achieved the right grades but not in the right subjects.
- The university or UCAS may not have received your results. The examination boards send the results automatically to UCAS, but if you sat an exam at a different centre, for example, then this may not have happened.
- The examination system that you sat does not automatically send the results to UCAS – for instance, if you sat overseas qualifications.

In cases of achieving the right grades but not in the right subjects, contact your firm choice university to discuss this with them. Universities may revise their offer and admit you if they still have places, or if you missed the grade by only a few marks, they may ask you to try for a remark. Exam boards change the marks in only a few cases, though, and they can go down or up, so don't place all your hopes on this. If you still do not receive an offer from your firm choice university and have not received an offer from your insurance choice university, then call your insurance choice university. If, by the end of this process, you still have no offers, you will need to enter UCAS Clearing.

UCAS Clearing - Tips

Clearing is the name given to the system in which all remaining course vacancies are advertised on the UCAS website and in national newspapers. In Clearing, you contact the universities directly that have publicised course vacancies and give them your grades and UCAS ID number. If you think that you might need to use the Clearing system, it is best to be well prepared because the vacancies are filled very quickly. Clearing is typically open from July to October.

Alongside their search tool, which includes over 30,000 course options, UCAS also offers Clearing Plus, a tool that 'matches' candidates to a list of courses in UCAS Hub. If you find yourself in Clearing, it is advisable to check the 'View matches' button; if you find a course you like, select the 'I'm interested' button. If the university or college still has available places, they will contact you to discuss further and possibly make you an offer.

Advice for Clearing

- Make sure that you have your UCAS ID number and a copy of your UCAS application ready for results day.
- Remain proactive! Use the Clearing Plus tool to speed up the process of finding another place.
- You need to have access to a phone that you can use exclusively, as you may need to make many calls over the course of results day.
- You also need to have access to the internet in order to access the directory of courses available through Clearing on the UCAS website. This is particularly useful as the website also has the university contact numbers that you will need to call.
- Think about the option of studying courses that might not be identical to the one you originally applied for, but are related. For example, sociology and psychology rather than single honours psychology.
- Be ready for an impromptu telephone interview. The admissions staff may ask why you want to study on the course, and you will need to have a little bit more tact than just saying, 'because I didn't get into the course I really wanted to'. Instead, you could say something like, 'Even though I didn't get in to my firm or insurance choices I did apply/intend to apply/visit during the open day/ know that the course has a good student satisfaction rating in the Guardian, etc.'

If you decide to retake your A levels

If you have not achieved the grades that you needed for your chosen universities and you do not want to take the available Clearing places, you could consider retaking one or more A levels. In the days when most examination boards offered January sittings, retaking might have meant studying for one term to boost the grade. The period from January to September could then be used to earn money, gain more work experience or travel the world. But, apart from international A levels, A level exams are now only available in June, and so retaking will involve studying for another year, so you need to be sure that your university aspirations are genuine enough to give you the motivation to add this extra year to your studies. As the A level system is now fully reformed, barring the last Phase Three legacy-subject examinations, you will need to retake the entire two-year qualification again and therefore plan to be able to do this in just a single year – you do not want a repeat of the examination if you are underprepared.

Speak to your teachers about the implications of retaking your exams. Some independent sixth form colleges provide specialist advice and teaching for students. Interviews to discuss this are free and carry no obligation to enrol in a course, so it is worth taking the time to talk to their staff before you embark on A level retakes. Many further education colleges also offer retake courses, and some schools will allow students to return to resit subjects, either as external examination candidates or by repeating a year.

If you decide to reapply

Universities are usually happy to consider students who are reapplying, either because they did not get the required grades the first time around or because they did not receive any offers of places. It is worth contacting the university to check whether this is the case. Some will have policies on grade requirements for retake candidates, while others might ask for evidence of any extenuating circumstances that may have affected the previous application.

> **TIP!**
> - If there were extenuating circumstances that affected your application, include a brief mention of this in the personal statement ('I was disappointed not to have achieved the required grades, because my studies were affected by illness, but this has made me even more determined to become an engineer'), but leave the details to the referee.
> - If you are retaking, you can use the extra term or extra year to add weight to your application, for example, by gaining more work experience, taking up a new subject, enrolling in evening classes that are relevant to your application and furthering your reading.

> *Engineering without imagination sinks to a trade.*
>
> Herbert Hoover, engineer

9 | Building bridges: Fees and funding

Whether undertaking an undergraduate or postgraduate course, the cost of studying is considerable. This has been exacerbated in recent years by rises in living costs due to inflation alongside the cost of university tuition fees. According to the UK Parliament's House of Commons Library, on average, students commencing their studies in 2022 were likely to accrue a student loan debt of £45,600. However, these figures were based on an average of all students, rather than focusing on engineering students. If we take into account the course length, then an engineering student could leave university with a significantly higher level of debt if the full amount of tuition fee and maintenance loan is received and there is no financial support from other sources.

When considering levels of student debt, it is easy to become disheartened and think that university study is not for you. What all students must remember is that tuition fees do not have to be paid upfront; in fact, most students receive student loans to cover this cost. In addition, the loans do not start to be paid back until you are earning over a certain amount . However, it can be a real challenge for many students to pay for living costs, such as rent and food, so it is vital to be aware of how you will meet these expenses.

Undertaking any university course should only be done after seriously considering the overall cost and carefully examining your ability to be fully committed to your studies.

Fees

UK

As each UK nation sets its own fees, the tuition fee that students have to pay for undergraduate courses will depend on where they live and where they intend to study. From 1 August 2025, the maximum annual tuition fee that providers will be allowed to charge will be £9,535, as part of the government's TEF, which assesses universities and colleges on the quality of their teaching.

9| Fees and Funding

There are a number of variations between the systems in England, Scotland, Wales and Northern Ireland, which can result in significant differences between the fees that are ultimately paid by students. In autumn 2024, the UK government announced that the tuition fee cap in England would be increasing from £9,250 to £9,535 for the 2025–6 academic year; this was followed by announcements by the Welsh and Scottish governments that they would bring their fees in line with England. Northern Ireland has increased tuition fees from £4,750 to £4,855 for 2025 entry. Therefore, the current rules are as follows, although they may be subject to change in the future.

- Students in England and Wales are required to pay a maximum of £9,535 if they are studying in England, Scotland, Wales or Northern Ireland.
- Students from Scotland who study at Scottish universities are not required to pay tuition fees (or, rather, tuition fees of £1,820 for 2025 entry are covered by the Student Awards Agency for Scotland [SAAS] for students who qualify for home student status). Scottish students have to pay fees of up to £9,535 if they study in England, Wales or Northern Ireland.
- Students living in Northern Ireland pay up to £4,855 if they attend university in Northern Ireland, and up to £9,535 if they study in England, Scotland or Wales.

EU and non-EU international students

At present, EU students are charged the same fees as non-EU international students, which are significantly higher than those charged to UK students and are determined by each university.

Some students from the EU may be eligible for support in terms of student loans from the UK government, but this is dependent on a number of factors, so it is best to check personal eligibility. Students from the Republic of Ireland are exempt from paying higher fees and are eligible for home fee status.

Living expenses

Your living expenses include the cost of your accommodation, food, clothes, travel and equipment, leisure and social activities – plus possible extras like field trips and study visits, if these aren't covered by the tuition fees.

Check university and college websites for information about possible living costs. Some are more informative than others and give breakdowns under various headings such as accommodation, food and daily travel. Others go even further and give typical weekly, monthly or annual spends.

If you're living away from home, accommodation will make up the largest proportion of your living costs. There is likely to be a range of accommodation options – from a standard room in university halls to privately rented accommodation – with a range of price points. You'll probably be surprised when you do some research to find that the cheapest and most expensive towns are not as you might have expected; the cost of accommodation often depends on how much of it is available in a particular area.

When choosing accommodation, it is essential to consider its location and factor in the cost of travel to your university or college. It is also important to find out what's included in the accommodation costs (such as utilities, personal property insurance and Wi-Fi) and whether it is possible to pay for accommodation during term time only.

Funding your studies

How do you fund your time in higher education? Don't ignore this question and leave it until the last minute! You will need to think carefully about how to budget for several years' costs – and you need to know what help you might get from:

- the government;
- your family or partner;
- paid part-time work;
- other sources, such as bursaries and scholarships.

This chapter gives a brief overview of a complicated funding situation, which can vary according to where you come from and where you plan to study. For more details about the different types of funding available and how to apply for them, check your regional student finance website.

- www.gov.uk/contact-student-finance-england
- www.saas.gov.uk
- www.studentfinanceni.co.uk
- www.studentfinancewales.co.uk

Tuition fee loans

For UK students, tuition fees can be covered by taking out a tuition fee loan, which will be paid directly to your university or college at the start of each year of your course. You are effectively given a loan by the government that you repay through your income tax from the April after you finish your course, but only once your earnings reach a certain threshold. Currently, these income thresholds stand at:

- £25,000 per year for students from England;
- £27,295 per year for students from Wales;
- £31,395 per year for students from Scotland (who go to university outside of Scotland);
- £24,990 per year for students from Northern Ireland.

So, if you never reach this threshold, you will not have to repay the fees. In addition, any outstanding balance on your loan will be cancelled after a certain period of time if you have not already cleared it in full. The length of time depends on the rules at the time you took out the loan. For students in England who started their studies after August 2023, the repayment period was extended to 40 years (from 30 years), so it is recommended that students in other regions keep a close eye on any developments with respect to the length of the loan repayment period. At the time of writing, the loan repayment term is 30 years for students from Wales and Scotland and 25 years for students from Northern Ireland.

The current situation regarding repayments is that you repay 9% of anything you earn over the annual income threshold.

The interest rate charged on student loans depends on which repayment plan you are on, but for students in England on Plan 5, it is currently set at 4.3%.

Maintenance loans

In addition to a tuition fee loan, all students can apply for a maintenance or living cost loan. All students are entitled to a maintenance loan, which is repayable in the same way. The amount you can borrow will be dependent on your household income – in other words, it is means-tested. 'Household income' refers to your family's gross annual income (their income before tax). With the exception of loans available to Scottish students, the amount you can claim also varies depending on your living situation, with the maximum loan being available to students living away from home in London.

Each regional student finance website includes a finance calculator tool that will give an estimate of the finance you would be eligible for based on your family income and other factors, and it is well worth looking at this before planning your budget.

England (2025-26)

The maximum annual maintenance loan in England is:

- £8,877 for those living in the family home;
- £10,544 for those living away from home (£13,762 in London).

Wales (2025-26)

In Wales, students can get a combination of a maintenance grant, which they do not have to pay back, and a repayable maintenance loan. Both are means-tested, but all students will get a grant of at least £1,000.

The combined total amounts available from maintenance loans and grants in Wales are:

- £10,480 for those living in the family home;
- £12,345 for those living away from home (£15,415 in London).

Scotland (2024-25)

In Scotland, all students can get a repayable maintenance loan, and those eligible will receive a non-repayable bursary (grant) to cover living expenses. These are as follows (all figures per year):

- household income up to £20,999: £2,000 bursary and £7,000 loan;
- household income £21,000–£23,999: £1,125 bursary and £7,000 loan;
- household income £24,000–£33,999: £500 bursary and £7,000 loan;
- household income £34,000 and above: no bursary and £6,000 loan.

Unlike the rest of the UK, household income for Scottish students is measured on the income bands listed above rather than exact household income.

From the 2024–25 academic year, an additional new 'special support' loan of £2,400 is available to all full-time students. Unlike the maintenance loan and bursary, this is not means-tested, but it is repayable.

Northern Ireland (2025-26)

The maximum annual maintenance loan in Northern Ireland is:

- £6,300 for those living in the family home;
- £8,132 for those living away from home (£11,391 in London).

In addition, you may be eligible for a non-repayable maintenance grant if your household income is below £41,065. This is paid alongside any maintenance loan you qualify for and is up to £3,475.

Sponsorship

Engineering students are more fortunate than their peers who are studying other subjects because of the large number of sponsorship and bursary schemes available from engineering institutes, companies and the universities themselves. This is because it is recognised that the UK needs to attract more able students into engineering. The starting points for finding out about sponsorship are:

- the university engineering departments;
- the professional engineering institutes (contact details are given in Chapter 11).

The professional engineering institutes or institutions are professional bodies that accredit and represent their members, provide training and information, promote their particular fields of engineering, organise or offer scholarships and help engineers with their careers.

The level of sponsorship varies from course to course, university to university, institute to institute and also changes from year to year. You will need to spend some time researching your options. The benefits of gaining sponsorship are numerous. In addition to financial assistance, you will have opportunities to gain internships, work experience or even a job to go to once you graduate. The very fact that an organisation or university has been prepared to sponsor you also says a good deal about your personal or academic qualities and will enhance your CV.

Sponsorship can include:

- financial aid during the degree course;
- paid or part-funded work placements during holidays or a gap year;
- work or study placements overseas;
- improved chances of getting jobs with the sponsoring companies after graduation.

Several publications giving details of the scholarships and bursaries offered by educational trusts are available, including the *Directory of Grant Making Trusts*.

Scholarships

There are a variety of scholarships available for engineering students – possibly more than for any other discipline. The professional engineering institutes also offer some scholarships – for example, the Institution of Mechanical Engineers, mentioned below. You can also use the website www.scholarship-search.org.uk to look for undergraduate and postgraduate scholarships.

University scholarships broadly fall under three categories:

1. scholarships offered by the university;
2. scholarships offered through donations by alumni;
3. industry scholarships, which can include work placements.

Normally, students apply for scholarships once they have accepted the offer of a place from the university, although sometimes universities will offer a scholarship alongside the offer of a place in order to attract the best students. There is a section on scholarships under the 'Fees and funding' heading on all university websites.

The Institution of Mechanical Engineers (IMechE)

Undergraduate scholarships and grants offer assistance to students who are about to start or have already started on mechanical engineering degrees accredited by the Institution. IMechE offers undergraduate scholarships worth up to £8,000 (£2,000 per year). Applications are open to UK students beginning their Engineering undergraduate degree courses. The application process is open until June and requires students to answer a Scholarship Award Question relevant to engineering. The 2024 question asked:

> *The current initiatives to reduce the single use plastic products will stem the flow into the oceans but the issue of cleaning up will remain. Discuss the key challenges to cleaning the oceans of plastic debris, describe what would you design to clean up the plastic debris and how your design would address the challenges.*

Details about specific scholarships and the application process can be found on the institution website (www.imeche.org/careers-education/scholarships-and-awards).

Special considerations

Extra help is available for students with a disability, mental health condition or specific learning difficulty, and for students with children or adult dependants. Students from all parts of the UK are also eligible for additional support from their university. This may take the form of note-takers and scribes for dyslexic students, funds for specialist equipment or additional tutoring. See the appropriate support website for further information on support grants and contact the student support department of your university to see what support it will provide.

Case study

Sasha is currently studying Civil Engineering at the University of Warwick. Having grown up in a family of civil engineers responsible for some of the biggest building projects of the past half-century, Sasha had always had an interest in complex projects. He was a normal student at IGCSE but particularly excelled in mathematics and physics, which he went on to study at A level.

It was through work experience that Sasha confirmed his higher education route, undertaking engineering experiences in various departments to make sure.

'Civil engineering appealed to me as I have always been interested in structures. I once thought that it was the design of them that I found interesting, but in truth, it is the complexity of the structures. Everything down to the depths of the foundations for the different buildings and how you can strengthen them even once the structures are standing. London opened my eyes to that. I have also been fascinated with history, and seeing how structures such as the Colosseum in Rome are still standing made me want to know how and why; we are still learning lessons from the past today.

'I would encourage students to explore as many different routes in engineering that they can before making a decision on their degree route because it is such a varied discipline.'

Did you know?

Theoretical physicist Richard Feynman (1918–88) used to crack vaults for fun in his spare time.

10 | A cog in the machine: Further training, qualifications and careers

There is no 'typical' day for a working engineer. One of the attractions of engineering as a career is that you are, with suitable planning, able to follow a career path that suits your own individual skills and ambitions. If you like working outdoors, working as an on-site civil engineer would allow you to do this, whereas if you enjoy working in a laboratory, then you might choose to be a structural or electronic engineer. As we saw in the introduction to this book, engineers can work as part of a large team for a multinational engineering company or on their own in their own company. Engineers who are interested in planning and finance can work as product engineers, assessing the economic viability of manufacturing a product and then designing the production line.

Some engineering graduates decide to change direction after graduating. For example, engineers are highly sought after in the financial sector because an engineering degree demonstrates that the student has analytical and problem-solving skills.

The engineering institutes (see list in Chapter 11) and university engineering departments provide a good starting point for further investigation of possible careers through the case histories that they publish. Almost all of the engineering organisations have sections on their websites (often under the 'Education' tab) that contain profiles of undergraduate and qualified engineers. Some good examples can be found:

- on the Royal Academy of Engineering's website: www.raeng.org.uk/education-and-skills/this-is-engineering;
- under 'Who are civil engineers' on the education pages of the Institution of Civil Engineers website: www.ice.org.uk/what-is-civil-engineering/civil-engineering-explained;
- at www.whynotchemeng.com (for chemical engineering) on the Tomorrow's Engineers website: www.tomorrowsengineers.org.uk.

10| Further Training, Qualifications and Careers

The websites of the university engineering departments are also useful sources of case histories, and they often have links that allow you to address questions to undergraduate engineers.

In its booklet *Educating Engineers for the 21st Century*, the Royal Academy of Engineering provides an overview of what qualities future engineers will need to possess:

> *No factor is more critical in underpinning the continuing health and vitality of any national economy than a strong supply of graduate engineers equipped with the understanding, attitudes and abilities necessary to apply their skills in business and other environments.*
>
> *Today, business environments increasingly require engineers who can design and deliver to customers not merely isolated products but complete solutions involving complex integrated systems.*
>
> *Increasingly they also demand the ability to work in globally dispersed teams across different time zones and cultures. The traditional disciplinary boundaries inherited from the 19th century are now being transgressed by new industries and disciplines, such as medical engineering and nanotechnology, which also involve the application of more recent engineering developments, most obviously the information and communication technologies. Meanwhile new products and services that would be impossible without the knowledge and skills of engineers – for instance the internet and mobile telephones – have become pervasive in our everyday life, especially for young people.*
>
> *Engineering businesses now seek engineers with abilities and attributes in two broad areas – technical understanding and enabling skills. The first of these comprises: a sound knowledge of disciplinary fundamentals; a strong grasp of mathematics;*
>
> *creativity and innovation; together with the ability to apply theory in practice. The second is the set of abilities that enable engineers to work effectively in a business environment: communication skills; team-working skills; and business awareness of the implications of engineering decisions and investments. It is this combination of understanding and skills that underpins the role that engineers now play in the business world, a role with three distinct, if interrelated, elements: that of the technical specialist imbued with expert knowledge; that of the integrator able to operate across boundaries in complex environments; and that of the change agent providing the creativity, innovation and leadership necessary to meet new challenges.*
>
> *Engineering today is characterised by both a rapidly increasing diversity of the demands made on engineers in their professional lives and the ubiquity of the products and services they provide.*

> Yet there is a growing concern that in the UK the education system responsible for producing new generations of engineers is failing to keep pace with the inherent dynamism of this situation and indeed with the increasing need for engineers.
>
> Source: www.raeng.org.uk/publications/
> reports/educating-engineers-21st-century.

While the publication is from 15 years ago, the ethos and sentiment have stood the test of time. However, this more recent statement from the Royal Academy of Engineering demonstrates quite clearly that there has never been a better time to try and enter the industry.

> There is now compelling evidence of a shortage of engineers in the economy . It is also clear that high-level technician skills are in particularly short supply and that future workers in engineering will require higher-level technical skills. The Review of Engineering Skills by Professor John Perkins FREng, Chief Scientific Adviser to the Department of Business, Innovation and Skills, has highlighted the complex skills pipeline into engineering jobs and access to, and progression through, the professional engineering qualifications.
>
> In a typical year cohort in England, 40% do not achieve a C grade or above in mathematics GCSE. Beyond compulsory education at age 16, the number of students choosing qualifications that lead to engineering careers plummets. The reasons for this are numerous, and include inadequate knowledge of engineering and poor perceptions and attitudes towards it, the perceived difficulty of subjects leading to engineering, the lack of specialist teachers in physics, mathematics and computing and the pressures of performance tables for schools.
>
> Source: https://raeng.org.uk/media/jdeb1ozi/
> the-universe-of-engineering-report-2014.pdf

Chartered Engineer status

Chartered Engineer (CEng) is a professional title registered by the Engineering Council. Engineers who achieve this status have been able to demonstrate that they have reached a high level of professional competence. Attaining the status of Chartered Engineer brings many benefits, including:

- being part of an elite group of highly qualified engineers;
- professional recognition of your qualifications and attainments;
- higher earnings potential;
- improved career prospects;

10| Further Training, Qualifications and Careers

- international recognition of your academic and professional qualifications;
- access to continuing professional training.

The normal route towards gaining the qualification is:

- an accredited bachelor's degree (BEng):
  This alone is not sufficient for CEng status but forms part of the pathway.
- Further learning to master's level. This can be achieved through:
  - an accredited integrated master's degree (MEng);
  - a separate accredited master's degree (MSc) following a BEng; or
  - equivalent further learning, as approved by the Engineering Council.
- Membership of one of the professional engineering institutes:
  Licensed institutions, such as the institution of Mechanical Engineering (IMechE) or the Institution of Civil Engineers (ICE), are responsible for assessing candidates.
- Experience of professional practice:
  You must demonstrate your professional competence and commitment through work-based experiences, typically recorded in a professional development framework and culminating in a Professional Review.

It usually takes 8–12 years from the start of an undergraduate degree to reach CEng status. For more details, contact the Engineering Council or one of the engineering institutes (see the list in Chapter 11).

Incorporated Engineer status

While this book focuses on you obtaining Chartered Engineering status (CEng), there is also Incorporated Engineer status (IEng), which we will mention here in brief.

The fundamentals of engineering form the base of both qualifications; however, there are key differences. The most common difference is that Chartered Engineers are considered to work towards developing solutions to engineering problems, whether that be in the creation or in the utilisation of new technologies. Incorporated Engineers are then responsible for the maintenance of applications and management.

The route that you embark on is generally determined by your employer and dependent on the skills and competencies they will require of you. Therefore, you should be careful when choosing your graduate scheme, so that you know what you are working towards.

If you wish to become an IEng, generally you would take a bachelor's degree first, though it is possible to become accredited in this

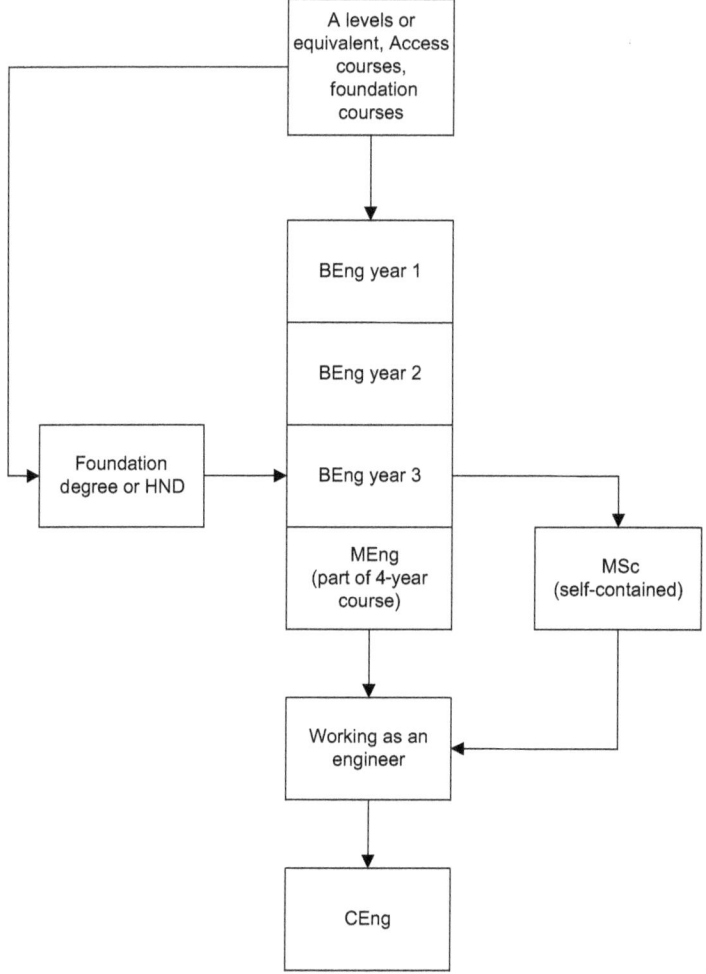

Figure 4 Routes to gaining an engineering qualification

qualification with a Foundation degree or HND and then further learning. You do not need to become an IEng before becoming a CEng. Although gaining a CEng may bring better pay compared with an IEng, choose your route based on what you are interested in rather than on financial reward.

Master's courses

Many undergraduate engineering courses are four years in length and lead to a master's qualification (MEng) rather than a bachelor's degree (BEng). The alternative route to a postgraduate qualification is by taking a self-contained MSc course after completing the bachelor's degree.

10| Further Training, Qualifications and Careers

Master's degrees allow students to focus their studies on a specific area of engineering. Applications for self-contained postgraduate courses are usually made directly to the universities, rather than through a central scheme. For listings of master's courses, see www.prospects.ac.uk and the university websites.

The advantages of doing a master's course are:

- greater specialisation;
- better job prospects;
- higher earning potential.

The cost of a master's course varies from university to university but is likely to be in the range of £8,950–£20,000 per year for tuition fees for domestic students and between £5,000 and £30,000 for international students. A number of grants and scholarships are available (see Chapter 9).

The structure of a master's course will depend on where and what you study. As an example, here is the current course programme for the MSc Energy Engineering with Environmental Management at the University of East Anglia:

- Energy Engineering Fundamentals;
- Energy Engineering Dissertation;
- Optional modules: Nuclear engineering and technology with advanced topics; advanced computational methods; energy infrastructure engineering; engineering group project; fluids engineering for renewable energy; solar energy engineering with advanced topics; and power systems engineering.

*Source: www.uea.ac.uk/course/postgraduate
/msc-energy-engineering.
Reprinted with the kind permission of the
University of East Anglia.*

Career opportunities and employment prospects

The engineering sector remains a cornerstone of the UK's economy, contributing significantly to economic output and employment. Recent data indicates that the engineering economy contributes up to an estimated £646 billion in direct Gross Value Added (GVA) annually, accounting for over 30% of the UK's total economic output. This sector provides high-value, highly productive jobs, with the average value of an individual engineering job at £70,000, nearly 25% more in GVA than the average UK job (Royal Academy of Engineering).

Employment prospects for engineering graduates continue to be robust. The sector employs approximately 8.1 million people, with 5.7 million in engineering roles and 2.4 million in non-engineering positions within engineering businesses (Royal Academy of Engineering).

The demand for engineering skills is projected to grow, with an increasing emphasis on green jobs and sustainable technologies. A recent report identifies approximately 6.1 million engineering jobs, including tech roles, highlighting the sector's expansion and the need for a skilled workforce to support this growth (Engineering Council).

In terms of remuneration, engineering graduates continue to enjoy competitive starting salaries. Research indicates that the average graduate salary in the UK is £35,170, though this figure can vary depending on the specific engineering discipline and employer (Luminate Prospects).

The engineering sector offers strong career opportunities, with promising employment prospects and competitive salaries for graduates. The industry's significant contribution to the UK's economy underscores the value and importance of engineering skills in driving economic growth and innovation.

11 | Light years ahead: Further information

Useful contacts

University applications

www.ucas.com

Funding

www.gov.uk
www.saas.gov.uk (Scotland)
www.slc.co.uk
www.studentfinancewales.co.uk (Wales)
www.studentfinanceni.co.uk (Northern Ireland)

News

www.bbc.co.uk/news
www.theguardian.com

Overseas voluntary projects relevant to engineering

www.ewb-uk.org/the-work

Organisations for engineers in the UK

Chartered Institute for IT
www.bcs.org

Chartered Institute of Plumbing and Heating Engineering (CIPHE)
www.ciphe.org.uk

Chartered Institution of Building Services Engineers (CIBSE)
www.cibse.org

Chartered Institution of Highways and Transportation (CIHT)
www.ciht.org.uk

Chartered Institution of Water and Environmental Management (CIWEM)
www.ciwem.org

Engineering Council
www.engc.org.uk

Institute of Acoustics (IOA)
www.ioa.org.uk

Institute of Cast Metals Engineers (ICME)
www.icme.org.uk

Institute of Electrical and Electronics Engineers (IEEE)
www.ieee-ukandireland.org

Institute of Highway Engineers (IHE)
www.theihe.org

Institute of Marine Engineering, Science and Technology (IMarEST)
www.imarest.org

Institute of Materials, Minerals and Mining (IoM3)
www.iom3.org

Institute of Measurement and Control (InstMC)
www.instmc.org

Institute of Physics (IOP)
www.iop.org

Institute of Physics and Engineering in Medicine (IPEM)
www.ipem.ac.uk

Institute of Water (IWO)
www.instituteofwater.org.uk

Institution of Agricultural Engineers (IAgrE)
www.iagre.org

Institution of Chemical Engineers (IChemE)
www.icheme.org

Institution of Civil Engineers (ICE)
www.ice.org.uk

Institution of Engineering Designers (IED)
www.institution-engineering-designers.org.uk

Institution of Engineering and Technology (IET)
www.theiet.org

Institution of Fire Engineers (IFE)
www.ife.org.uk

Institution of Gas Engineers and Managers (IGEM)
www.igem.org.uk

Institution of Mechanical Engineers (IMechE)
www.imeche.org

11| Further Information

Institution of Power Engineers (IPowerE)
www.ipowere.org

Institution of Royal Engineers (InstRE)
www.instre.org

Institution of Structural Engineers (IStructE)
www.istructe.org

Nuclear Institute (NI)
www.nuclearinst.com

Royal Academy of Engineering
www.raeng.org.uk

Royal Aeronautical Society (RAeS)
www.aerosociety.com

Royal Institution of Naval Architects (RINA)
www.rina.org.uk

Society for the Environment (SocEnv)
https://socenv.org.uk/page/see-support
Society of Operations Engineers (SOE)
www.soe.org.uk

The Welding Institute (TWI)
www.theweldinginstitute.com

Specialist publications

Architect magazine
www.architectmagazine.com

Aviation Week
www.aviationweek.com

E&T [Engineering and Technology] magazine
https://eandt.theiet.org
www.theengineer.co.uk

Engineering News Record
www.enr.com

Nano
www.nanomagazine.co.uk

New Civil Engineer
www.newcivilengineer.com

Race Car Engineering
Electronic Engineering Times
www.eetimes.com

137

The Engineer
www.theengineer.co.uk

Engineering News Record
www.enr.com

Nano
www.nanomagazine.co.uk

New Civil Engineer
www.newcivilengineer.com

Race Car Engineering
www.racecar-engineering.com

Structure magazine
www.structuremag.org

Books

Engineering

- Blockley, David, *Bridges: The Science and Art of the World's Most Inspiring Structures*, OUP, 2010.
- Brain, Marshall, *The Engineering Book: From the Catapult to the Curiosity Rover*, Sterling, 2015.
- Brenner, Brian, ed., *Don't Throw This Away!: The Civil Engineering Life*, American Society of Civil Engineers, 2006.
- Browne, John, *Seven Elements That Have Changed The World: Iron, Carbon, Gold, Silver, Uranium, Titanium, Silicon*, Weidenfeld and Nicholson, 2013.
- Dupre, Judith, *Skyscrapers: A History of the World's Most Extraordinary Buildings*, Black Dog & Leventhal Publishers Inc., 2008.
- Dyson, James, *Against the Odds: An Autobiography*, Orion, 1997.
- Eberhart, Mark E., *Why Things Break: Understanding the World by the Way it Comes Apart*, Three Rivers Press, 2004.
- Fawcett, Bill, *It Looked Good on Paper: Bizarre Inventions, Design Disasters and Engineering Follies*, Harper Paperbacks, 2009.
- Gordon, J.E., *Structures: Or Why Things Don't Fall Down*, DaCapo Press, 2003.
- Gordon, J.E., *The New Science of Strong Materials: Or Why You Don't Fall Through The Floor*, Penguin, 1991.
- Hart-Davis, Adam, *Engineers*, Dorling Kindersley, 2012.
- Michell, Tony, *Samsung Electronics and the Struggle for Leadership of the Electronics Industry*, John Wiley & Sons, 2010.
- Munroe, Randall, *Thing Explainer: Complicated Stuff in Simple Words*, John Murray, 2015.

- Monroe, Randall, *What If: Serious Scientific Answers to Absurd Hypothetical Questions*, John Murray, 2015.
- Open University (author), *Engineering: the Nature of Problems*, Open University, 2016.
- Petroski, Henry, *Invention by Design: How Engineers Get From Thought to Thing*, Harvard University Press, 1998.
- Sammartino McPherson, Stephanie, *Tim Berners-Lee: Inventor of the World Wide Web*, Twenty-first Century Books, 2009.

Thinking skills

- Butterworth, John and Thwaites, Geoff, *Thinking Skills*, CUP, 2005.
- Tanna, Minesh, *Think You Can Think?*, Oxbridge Applications, 2011.

> **Joke**
>
> Customer: Do you have any 2-watt, 4-volt bulbs?
> Sales Rep: For what?
> Customer: No, 2.
> Sales Rep: Two what?
> Customer: Yes.
> Sales Rep: No.

Glossary

UCAS applications

Admissions tutor
Someone within an engineering department who deals with UCAS applications.

Clearing
The period from early July to mid-September when students who are not holding offers for undergraduate university places can approach universities that still have vacancies.

Deferred entry
Applications for an undergraduate place for the following year, allowing the student to take a gap year.

Extra
Allows students who are not holding any offers to approach extra universities, prior to receiving their examination results.

Gap year
A year between leaving school or college and starting university, usually used to gain further work or life experience or extra qualifications.

IELTS
The International English Language Testing System. Students who do not have English as their first language must reach a certain IELTS level in order to gain entry to study in the UK.

UCAS
The Universities and Colleges Admissions Service, the online undergraduate application system.

Engineering

Aerospace engineering
A specialist branch of mechanical engineering focusing on aviation.

Artificial Intelligence
The theory and development of computer systems able to perform tasks normally requiring human intelligence.

Glossary

Automotive engineering
A specialist branch of mechanical engineering dealing with transport.

BEng
The qualification gained after a three-year undergraduate engineering degree course (often four years in Scotland).

Biomedical engineering
Linking engineering and living things.

Chemical engineering
The branch of engineering that looks at industrial processes involving chemicals, drugs, food and fuels.

Civil engineering
The branch of engineering dealing with large-scale infrastructure projects such as roads, bridges and dams.

Electrical engineering
Engineering involving electrical devices. Often taught as a joint degree with electronic engineering.

Electronic engineering
The branch of engineering that deals with electronics, such as integrated circuits.

Energy engineering
Engineering associated with energy production and transmission, environmental issues and alternative energy sources.

Mechanical engineering
The branch of engineering dealing with machinery.

MEng
The qualification gained from a four-year engineering degree course (often five years in Scotland).

MSc
A self-contained master's postgraduate degree.

NRA
National Risk Assessment.

Production engineering
Engineering that relates to manufacturing processes.

Structural engineering
Engineering that deals with the use of suitable materials for engineering projects.

Have you seen the Getting into University series?

Get 30% off when you buy the complete series!

Written by experts in a clear and concise format, these guides go beyond the official publications to give you practical advice on how to secure a place on the university course of your choice.

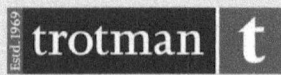

Order today from
www.trotman.co.uk/GettingInto

www.ingramcontent.com/pod-product-compliance
Lightning Source LLC
Chambersburg PA
CBHW042137160426
43200CB00019B/2963